A WESTMORLAND

His life, poems and songs

By Fred Nevinson

Compiled by Anne Bonney

Aided by the Curwen Archive Trust

*Dedicated to
my dear wife and family.*

Published by Helm Press
10 Abbey Gardens Natland Kendal Cumbria LA9 7SP
Tel: 015395 61321

Copyright - Anne Bonney

First Published 1997
First Reprint 1997
Revised Edition 2000

Typeset in Galliard 9 & 10 pt

ISBN 0 9531836 4 5

Typeset and printed by Miller Turner Printers Ltd
The Sidings Beezon Fields Kendal Cumbria LA9 6BL Tel: 01539 740937

Cover: Fred at Borrowdale Head, August 1997
Photo by Phillip Bonney

CONTENTS

Page No.

Introduction		2
Where I have live and worked		3
1.	Childhood days in Little Langdale	5
2.	Working round Ambleside	23
3.	My years at Selside and Ravonstonedale	27
4.	Bannisdale Head and back to Selside	41
5.	Borrowdale Head to my retirement	49
6.	Merry Neets	61
Poems and songs		
	Solitude	62
	Lakeland	63
	The Langdales	64
	Mountain Glory	64
	An Old Man's Tribute to Lakeland	66
	The Lakeland Rambler	67
	The Little Langdale Lad	68
	The A6 Road	70
	Borrowdale Hunt	73
	Our Johnny	74
	A Tribute to John	76
	Anthony Barker's Hare Hounds	78
	A Tribute to a Huntsman	80
	That Cunning Old Fox	81
	Borrowdale Hunt 1981	83
	Greenholme Hunt	84
	Longsleddale Hunt (Fictional)	86
	The Hunters	88
	Old Darky	89
	Memories of Willie Porter	91
Epilogue - Farewell		93

INTRODUCTION

I have just had my birthday, eighty-one years, everything has changed a lot since I was born. I am trying to remember as much as I can and put it down for posterity.

Since we retired, we have done a lot of fell walking, taking photographs and have some two thousand slides of Lakeland animals and flowers, have shown them often. I have had five holidays, one in Cornwall, one at Northampton and two at Melton Mowbray and my honeymoon. I have not travelled far, but I know the district well where I live and love. I've been on a train three times.

We moved to Kendal in 1988, it was a wrench, nearly broke my heart. We celebrated our golden wedding in 1989 and had a good time at home. We are failing now, Maggie had a stroke three years ago, I got diabetes at seventy-two, had operations on both legs, we have had a good life together. The Lord has been good to me, I have a good wife, three caring family, four grandsons, all good lads and healthy, four great grandchildren, all well. God bless them all.

People who read this may think I've led a dull life, but that is not so, it's been great enjoying the simple things and living with nature. Feeling as I do now, I would like to go through it all again. I would like to thank all the good friends I have made in my life, for the pleasure and kindness they have given me and sorry to say, many have now passed on. I lived my life as I wanted and other people can live theirs as they like, I'm not bothered how they do it. Writing this has been much harder than I thought it would be. I will have missed some things I should have wrote, perhaps wrote some I should have left out. I have tried to make my story as accurate as possible but if I have made any mistake, I hope my readers will excuse me. I think the next event for me will be my funeral. Everything I have said is true, this is not fiction.

I wish to thank Anne Bonney, John Marsh and the Archives Department of the Cumbria County Council, in particular, Richard Hall, for without their help this book could not have been completed. I, in particular, appreciate the generous grant that the Curwen Archive Trust have given to assist in the publication of this book.

Fred Nevinson
Summer 1997

WHERE I HAVE LIVED AND WORKED

Skelwith Fold, near Skelwith Bridge, Ambleside (Born 27.5.16) - moved short while later

High Park, Little Langdale - until 15 years old

Round Hill Farm, Ambleside - 1931 for approximately 2 years

Grove Farm Ambleside - 1933 and stayed 1 year

South Gateside, Selside - came in November 1934 and stayed 2 years

Forest Hall, Selside - 6 months

South Gateside, Selside - 6 months

Bainsbank Farm, Middleton, Nr Kirkby Lonsdale - 6 months

Above Park, Selside - 1936 for approximately 3 years until October 1939

Park House, Ravonstonedale - approximately 3 years

Bannisdale Head, Selside - from October 1943 until June 1954

Above Park, Selside - summer 1954 for about 2 years

Borrowdale Head, near Shap - 1956 until after retirement in 1988

Kendal - from 1988 to present day

Little Langdale School 1921

Back row, left - Turb Barrow, Joe Youdell, Hilda Walker, Pauline McCabe, Percy Rawes and Jack Barrow.

Third row, left - Billy Atkinson, Lizzy Rawes, Jennie Leck, Phyllis Mackintosh, Beattie Nottal, Dot Fell and Ernest Hodgson.

Second row, left - Annie Clark, Edith Hodgson, Kathy Parker, Mollie Birkett, Dorothy Hodgson, Corena Graves, Patty Parker, Mollie Nevinson and Betty Graves.

Front row, left - Fred Nevinson, George Hodgson and Jim Birkett.

CHAPTER ONE

Childhood Days in Little Langdale

I was born at Skelwith Fold, on 27 May 1916, a little hamlet between Hawkshead and Skelwith Bridge. My forelders lived at Newby Hall, near Penrith, and the Nevinson coat of arms are on Newby Hall. The family looked after the Yeomanry and horses for the Lonsdales in the 17th century. When the Lonsdales moved and opened the pits at Whitehaven, the Nevinsons moved with them. My great grandparents and families drifted back this way. I have an old, very distant relation, Harry Nevinson, who lives at Kirkland, near Ennerdale, and he has gone to Churches and Archives to trace the name back through many years.

When I was a fortnight old, my dad, William Hodgson Nevinson, born on 22 February 1884, went into the army. My mother, Annie Mary Nevinson, nee Bowness, born on 18 January 1887, moved with me and my three year old sister, Molly, to High Park, Little Langdale, into a cottage near her parents' farm. Granddad, George Bowness and grandmother Anne, ran the farm with the help of one of his sons, called Tom. Granddad was also a blacksmith and he shoed horses and sharpened quarry tools. We got our milk from him, he had a blue-grey cow he saved the milk off her for me and I was reared on it. I was only on a bottle for a week, then a cup, up to about five years old I always sucked my food instead of chewing it. Someone said, "It was with me not sucking when I was a baby." Sometimes when I got into mischief when I was older, granddad would say, "We should of drowned thee and reared a calf on the milk." This was raw untreated milk not like it is today. I was above average weight at a year old.

Granddad had a horse and trap, which he and Uncle Tom ran for hire and took me to Brathay Church in 1916, three and a half miles each way, to be christened. This was in early June a fortnight after I was born and before dad went off to war. Babies were christened early in those days. I was two years old when dad came home from the war. When I was three to four years old, I had a head of golden curls. There was a lady artist who wanted to paint my portrait but I would not keep still. Her name was, Miss Parnell, and she built a bungalow near Park Farm, at Skelwith.

As I said, mother's brother, Tom Bowness, worked on the farm and I spent a lot of time with him. I adored him and he felt the same about me, he was a dad to me. I don't remember my father's dad, his name was John Nevinson, and his wife was called Hannah. They lived in Windermere after they retired. I was only very young at the time. He had been a quarry man and worked at Little Langdale. It was too far to travel to go and see them but dad would go on his bike to visit. Grandma came to visit odd times but not very often and I don't know how she travelled, but that was after granddad died and she came on her own. She died about the same time as I started school in the 1920's.

My sister, Molly, started school at Little Langdale and I was on my own. I had a cousin, called Alf Hodgson, he lived half a mile away and he was four months younger than me. We would meet up and wander about, they would sometimes find us lying asleep under a hedge. I don't remember all our exploits, we were only four years old then but my

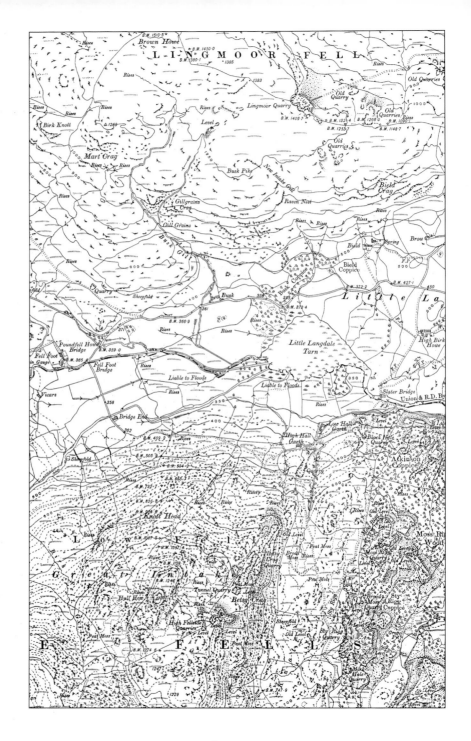

Reproduced from 1920 Ordnance Survey Map.

7

sister told me some. My most vivid memory of that time was when my younger sister, Dorothy, was born. She was born at home, it was the custom then. Mrs Allonby was the midwife in the dale. I was sent to play with Alf and wasn't allowed home till night. My sister took me upstairs, mother was in bed with the baby, I remember it as if it was last week. Later my sister, Dorothy, poor lass, died with cancer aged forty-six, a bitter blow to us all.

Alf moved further away and I spent more time with Uncle Tom. I was with him ploughing one day, walking too near the plough, when it hit a rock, the stilt (plough handle), came up and hit me on the head, knocked me out for a few seconds. I think it affected me for some time.

I remember going to Coniston Railway Station with Uncle Tom to get our coal and he would cart coal for others that wanted it delivered. We used a lot of wood on our fires as well, it didn't cost anything, the coal did!

I started Little Langdale School when I was five years old. I had a mile and half to go to school. We had two teachers, the infant teacher was called Miss Titterington, she was a fine well developed young woman and was a very happy go lucky person. Us boys called her 'Miss Tit' for short. The name was appropriate we thought, valley lads were not all backward!

Miss Wolfe taught the senior children and she could handle a cane. At one time there were seventy-two children going to school, there were a lot of large families and quarry men's children. We started off with a slate and slate pencil and used a piece of cloth to remove our work. We moved on to paper and pencil, then pen and ink. We had two at a desk, the lid would go up and we would keep our books and pencils inside. Some of the old desks still had four or five sitting at them.

There were only two classrooms with a large panel dividing them, which could slide open and make into one large room. In winter they lit the two metal stoves to try to keep the classrooms warm, though it was still chilly. We had our separate playgrounds for the girls and boys, with a wall dividing us and toilets at the bottom for each. When the bell rang the boys lined up at one end of the school and the girls at the other. Most of us had clogs on and we made quite a racket on the wooden floorboards and these became uneven with knots exposed.

There were separate cloakrooms also for hanging our coats in. We had prayers every morning. We did writing, reading, arithmetic and drill (now known as gym). The teachers were good and those wanting to, were taught music in the evenings after school or at weekends. I'm afraid I did not do too well at school, my mind was always with Uncle Tom, the animals and wildlife. As I grew older, I ran about the hills like a young deer!

We took packed lunch to school, which was usually jam sandwiches and cake, with cold tea in a bottle, water or lemonade to drink and in winter the teacher would make us hot cocoa. The school has been closed nearly twenty years now and is used as a Holiday Centre for schools to come and enjoy Lake District pursuits from all over the country.

High Park, Little Langdale - where I spent the first 15 years of my life.

Interior of Little Langdale School about 1900.

John Marsh Collection.

My Dad in his early 30's, on the left, with a friend in army uniform.

We were allowed out at dinner break and often some of the bigger boys, I was getting to be one, would play fox and hounds. We were all hunters, it was our entertainment. I was always chosen to be the fox, I could run the fastest. I would put out in a circle and if I got back to the school yard before they caught me, I'd gone to ground! They did not catch me very often. I had always just a few minutes' start.

Going back to when I was younger and dad came home from the war in 1918. I had been spoiled by my mother and Uncle Tom. Mother was kind and gentle, dad was the opposite, he started knocking me into shape and I mean knocking. I hated this strange man, I disliked him until I left school. He was a very good scholar, he was talented, he could draw, paint and was good at wood carving. I had none of his skills, I think I was a disappointment to him. He had a sideboard made up of wood carvings that he had done. The two drawer panels were of a weasel, bird and ivy, the two doors are of a sparrow hawk and heron, with ferns on the overmantle. Dad had seen these wildlife scenes, came back home and drew them first before carving. A friend of his who was a joiner, made them up into a sideboard, which I proudly have in the sitting room to this day. He also carved an overmantle which was given as a wedding present to my mother's oldest brother and my son Brian now has this. He also carved picture frames among other things.

I can remember dad was in the Machine Gun Corps, in France and he told me the story of when he was under this road, where the water came through a culvert and he was in there out of sight when they were shooting, he must of been outnumbered or something. He was there a long time and he got very wet and cold. Hypothermia, it would be called now. His legs were dead up to his hips and I can't remember how he got out. He was sent to hospital because he couldn't walk and he was transferred back to Britain, to Grantham. I can't remember the name of the hospital. When he was pulling me up later for mischief, I always said, "You go back to Grantham." I was only about three years old at the time when I was saying this.

When I was older, he would help me with my school work and show me other things to do that he was good at, like drawing. If I was not doing very well he would clout my ear, call me useless and leave me. This made me worse. To this day, I hate anyone standing and watching me doing a job. As I grew older, I could understand him. He'd had a rough time in the war and couldn't get work in the quarry when he came home and that was his job. I was always with Uncle Tom and this did not help matters.

My cousin Alf and I were always good pals all our school days. In summer, we would go to the fell gate on the Fell Foot road, there is a cattle grid there now. Sometimes there were two coaches, sometimes three or four. A coach with two horses was called a two-in-hand, three horses, a three-in-hand and four, a four-in-hand. We would open the gate when the coach came, the folks would throw coppers down for us. We sometimes got six or eight pence each. It was a lot then, but we had to hand it over when we got home and we would walk past the sweet shop and not spend a penny! It was the thrill of just having that money inside our pockets for an hour! Sometimes there was a trip, four coaches altogether, they would go full trot and someone would be blowing the post horn, it was lovely! I would think there could still be some coppers in the gutters and rushes nearby, which we never found!

The coach firms did not have work for their horses in winter, so they let them out to ploughing farms in the Furness district to work for their keep. One Sunday I remember in May, we were out walking on the road above Skelwith, when we heard the sound of galloping horses. They came past us, one man had five, another man had six, they were riding one horse, with another coupled to it. The pair behind had their halters tied to the tails of the horses in front. They had no time to kick one another! They had been collected to go back to their summer coach work. It was quite a sight!

Once a fortnight we went to Church, which was Church of England, but we just called it Chapel and it was joined onto the school, but there was no Sunday School. Services were in the afternoon, after the vicar had preached at Great Langdale. We had a harmonium for music.

I still spent a lot of time with Uncle Tom, he taught me how to fish and catch moles. The farmers in the dale did not keep many cows, not enough to keep a bull. Another uncle of mine, on dad's side, Johnny Youdell, did keep a bull and farmers took their cows there. The fee was five shillings. I think this custom invented the saying 'Parish Bull'. I often helped Uncle Tom to drive his cows to visit their boyfriend. When we got there, Aunt Mary would shout, "Come in Fred for a piece of cake!" I was not allowed to watch the performance!

I think it was about 1926 when the first cars came to Langdale. The Landlord, William Dixon, of the 'Tourists' Rest' (now called 'The Three Shires'), had one of the first cars in the valley, it was a black 'Model T' Tourer. Also at this time, there was a young man called, Horace Pepper, who was brought up by two maiden aunts, called the Miss Peppers. He was spoiled and they bought him a 'Model T' Tourer, this one was maroon, with a canvas top. He was a good sport and when he went passed the school to turn at the road ends, if it was playtime or dinner time, he would load as many of us into the car as possible and we would ride to the road ends and back about three hundred yards, it was a great thrill!

Both men set themselves up using their cars as taxis and there was a certain amount of rivalry between the two. I remember that soon after the two cars had come to Langdale, I was in a field with Uncle Tom overlooking the road looking up the dale, where we could see the road for half a mile or more. We saw Mr Dixon and Horace's cars were travelling towards one another from opposite directions and I don't think they knew the other was coming and from where we were stood, we soon realised they were going to meet on a blind corner. They were the only two cars for miles around and they had a bump, fortunately it was just a small one! They had only been going about 25 mph, the cars were strong and no damage was caused. Uncle Tom said, "I wonder what they are saying to each another?" Horace later married, moved away and he and his wife gave ballroom dancing lessons at Morecambe, I think.

About this time the first telephone was installed in the valley, it was at the 'Tourists' Rest' and everyone paid so much towards the cost of it. They came round and collected so much for putting the wires in. When Jack Robinson came round canvassing for the subscription for the telephone, I remember him saying to granddad, "If your wife goes on holiday (which was highly unlikely in those days), if she runs short of money, you can send it down the phone!" I never used it because I was too young but people would go with their message and Mr Dixon would do it. With it being new, some of them were a bit frightened to use it.

Also at this time, some parents formed a Committee, they had whist drives and dances to raise funds, which they used to take the children on a trip to Morecambe every year. There was a meal and half a crown to spend (2s.6d equivalent to 12½p today), it was a great treat! All the parents paid their own fare and the cost of their meal. The bus was usually hired from Brown's of Ambleside, who were later taken over by Faulkener's but kept the name of Brown's. Bells from Ambleside, also did coaches but they were mainly horse drawn.

The coach would pick us up about 7.30 am at Colwith, at the bottom of Langdale. There was much excitement. The bus was a single decker with an open top, which had a canvas hood that could be rolled back. I can't remember how long it took to get to Morecambe, but the bus would not be going fast. Morecambe we thought was great! We would be planning for weeks what we were going to spend our money on, half a crown was a lot of money then and could go a long way. We would go to the amusement arcades, where there were the usual amusements in these days, like coconut shies, duck shooting, penny rolling and roundabouts. We used to play in the sand, paddle in the sea and have donkey rides. We always looked forward to our tea, which was part of the treat and we were all booked into a cafe. All the children sat together and the adults separately. Quite often our leaving time was delayed by a lost child but we usually arrived back in Langdale about 7.30 pm, and as often as not, I had some money left over in my pocket. We all slept well that night!

Uncle Tom got married about this time, to Mary Birkett, from Side House Farm, at Great Langdale and I had to take a back seat. She was a lovely, very kind person.

Another game we played was at Bield Borran, when we were eight or nine years old and there was a fox's earth made out of boulders, not earth and we would go in at the bottom for about twenty to twenty-five yards. This was just for the sake of something to do when playing or rambling about. We went in with our clogs on and clattered about. One day one of us met a fox and we soon bolted. I don't know who was more scared, us or the fox!

We also used to play games whereby we were either quarrying or farming and we used to get hammers and try to break rocks. We got the rocks that were rotten and tried to break some pieces off. When we were playing at farming, this was a game I usually played at home with Alf. We used the fir cones of the fir trees for animals. Larch was for Herdwick sheep, scots pine were Swaledales and the big long cones of the spruce trees were cattle. We had to as the saying goes, 'make do and mend', we couldn't afford real toy animals. We used to put stones in a circle on the grass to make fields. I saved cigarette cards, pictures of footballers and cricketers and still have some of the sets yet.
On Wednesdays it was games at school and when I was about eleven to twelve years old, I remember we would sometimes have a paperchase. First we tore paper into pieces, put

A quarry man working on the face of Thrang Quarry, Chapel Stile.
This was one of the jobs that my Dad did.

John Marsh Collection.

Uncle Tom with his horse and cart at Coniston Railway Station, which he used for hire, waiting for a passenger. Note the candle lamp on the trap (this was a thick 'trap candle').

Grandmother Nevinson on the right, Aunt Mary Youdell on the left and cousin Ivy Burton, going home to Windermere after being on holiday at Birk Howe, Little Langdale, in Horace Pepper's 'Model T' Tourer.

it in a satchel and scattered it in handfuls as we ran. Two went first with the paper and the others followed the trail, it was good exercise.

Mrs Pepper lived at Greenbank, which was a small holding. Her husband was called, Jack Pepper, and he was the 'odd job man', doing soldering, joinery and repairing things. He could turn his hand to almost anything. There was no welding then. She always said, she knew when Nevinsons were going to school on a sunny morning, as she could see our clogs shining. They were always highly polished, mother would do them.

I remember at this time too, we used to have a fish man called Mr Kalley (he was a flat race runner at sports' days), he came to Little Langdale by bus from Kendal. I can't remember how often he came but he would have two baskets of fish, containing finney haddock, bloaters and kippers, and would go up one side of the dale and come down the other. We would be the last house he would call at before catching the bus home and mum always gave him a drink of tea and a pasty, in return he always gave mum what fish he had left, rather than take it home. We enjoyed the treat!

Our doctor at this time was Doctor Kendal. He was a big strong man, with a white beard and would go on his rounds on horse back, horse and gig, or bike. My father told me a story about him. He said that the doctor had been to the Duddon Valley delivering a baby and that he was making his way home to Coniston. It was dark and he was coming over Wrynose Pass, the road on the top of the fell was grass, as it was not used much then. He could not hear where the horse's hooves were treading and it was getting hard to guide. His horse stopped and refused to move. The doctor got off and struck a match, he could see why the horse had stopped. It had sensed danger and had stopped on the edge of a rock and there was a twenty foot drop. He gratefully turned the horse round, mounted and gave the horse a loose rein and it rejoined the track for the safe homeward journey.

Most of the men in Little Langdale worked in the quarries and the quarry owners built two rows of houses in Little Langdale. One was called Fitz Steps and the other Green Bank. They were let to their men for a small rent. Fitz Steps was also known locally as 'Jam Street', as the quarry men always had jam sandwiches in their bait (lunch) boxes.

Some men walked over two miles to work, often the last half mile was up a steep mountain. The drilling for blasting was done by hand, often slung sitting on a rope, working on the actual rock face some way up from the rock bottom. Some men were paid a small wage, while others were paid per load. When there was a lot of unemployment, after the war, the men that used to be quarry men and could not get work, often tried to find good slate to sell, sifting through broken and waste, to find a good seam that had been left in old disused quarries. If they were unlucky they moved on and others would maybe come along and try their luck.

High Fell Quarry was high above Tilberthwaite, the slate was brought down by horse and cart and sledge, it was a steep track. The cart was taken about half way up, then the sledge was taken to the top and filled and loaded onto the cart and repeated. This load was put in the cart with the first sledge load. A half sledge load was then fastened to the cart axle with a chain and this acted as a brake for the cart and eased the weight on the horse's back. The slate was used by local builders and was also taken to Coniston Railway Station, where it was transported to other parts of the country.

I knew men who lost their lives in the quarries. Two were killed in Hodge Close Quarry, they had gone back inside the cave too early after the explosion and some rock fell on them. Another time, was when one quarry man was in the open quarry at Elterwater, this again was also after an explosion, when he went in to knock down a piece of rock that was hanging and the rock that he was standing on gave way. All the quarry cottages are now holiday houses or weekend homes and the old valley life has gone.

There was a cattle dealer, I remember from West Cumberland and he was called, Henry Watson. I can still picture him now, a big tall man in neat cut breeches, brown shoes and leggings. Mr Watson came to Coniston on the train and walked the district buying cattle for about a week until he had all he wanted. He had his places where he stayed the night, usually he stayed one night at the 'Tourists' Rest' and other nights at local farms before going back home the same way he had arrived. There were no cattle wagons then. After he left, the people at the head of the valley would start to drive their cattle that Mr Watson had bought down the valley, others would join in as they passed and they would all meet at the bottom of the valley. There would be about thirty head of cattle of various sizes, with three or four people helping, one would be in front, two going either side and one behind. They travelled quietly along the quiet roads, dogs were not used in case they frightened them. One lad I knew, his cattle were just like pets, so when he went in front they followed him like a dog and the rest followed on. From the valley, they would continue to drive them along the Coniston to Ambleside road, to Coniston Railway Station. The whole journey would take a couple of hours. They were then loaded on the train and taken to first Barrow, then round the coast to Whitehaven or Workington. From there Mr Watson would take them to auctions at Cockermouth, Wigton and West Cumberland. The Langdale farmers sold their cattle to Mr Watson, as they were a long way from markets at Kendal, Ulverston and Broughton. Another important fact being that they had no carriage or commission to pay.

Other farmers would take their cattle and sheep to Ambleside Fair. This was held twice a year, the first and last week in October. This was just livestock, the majority being sheep and some cattle. The sheep were mainly Herdwick and some Wensleydale. The cattle were Shorthorns, Galloways and a few Highland. The Fair was set up in a field in the outskirts of Ambleside with temporary pens and the livestock was brought in from the surrounding valleys.

The first bus we had, was Creightons from Ulverston, and that made three trips a day. It travelled from Ulverston to Ambleside and we caught it at Oxenfell to travel to Ambleside. It was ninepence halfpenny return, three and a half miles each way.

The butcher came once a week from Grasmere with a sort of trap, it had a seat for the driver and a big wooden container. The back let down and on it was a chopping board, with weights hanging from the roof. His name was Tom Chew, a good name for a butcher!

A traditional hay making scene in Forge Field, Little Langdale, in the 1950's.
Left to right - John Birkett (George's son), Ted Nevinson(my uncle), Jim Hodgson (my brother-in-law) and George Birkett (my brother-in-law).

John Marsh Collection.

The 'Tourists' Rest' in the early 1900's later known as the 'The Three Shires'.

John Marsh Collection.

Eskdale Foxhounds coming up Bield Hill, Little Langdale, Arthur Irving in front and Jim Hodgson (my brother-in-law).

I remember some of the old characters, not to mention any names. One used to tell tall stories -

"He was out shooting and saw a rabbit sitting across the beck, he lifted his gun to fire. As he did, a snipe flew past, a fish rose from the water and he shot all three."

"Two were arguing in the pub and one had lost an eye at the quarry. He did not have a glass eye, just the lid over the hole. His opponent called him, "A blind old b......" He said, "I can see more than thee, I can see thy two eyes and thou can only see one of mine.""

One chap was farming with his mother. He was to be married in July. It so happened there was a heatwave at the time of the wedding. He went out of the hay field in his shirt sleeves, had the service, then back to the hay. I think he had his honeymoon when the weather broke! The marriage lasted fifty years.

There were eight in mother's family, most people married local. Two of mother's brothers married two Allonby sisters, one brother and one sister married Dixons, another brother and sister married Hodgsons. It was said that, if you kicked one, they all limped. The Hodgsons were on Busk Farm and the others weren't farmers. Mr Allonby was a woodman and his wife was a midwife. The Dixons were the same family that had the 'Tourists' Rest'.

Uncle Tom had a shot gun but he did not use it much. My dad often borrowed it to shoot vermin. He was a very good shot. He used to heat oatmeal on a shovel till it had a good smell. He put it near the rat runs and they would soon come out. I've seen him get six or seven with one cartridge. He came home one night and said, "I got two carrion crows with one shot, they were mating on top of a fir tree!" Mother said, "That was an unkind thing to do." He said, "Oh, they died happy!" One of his rare humorous moods.

I remember we paid half a crown a month into what was known as the 'Doctor's Club', this covered the whole family and I think somebody was employed to come and collect it. This was of course pre National Health days but we more or less treated ourselves, it was always bottles of medicine then, there was no pills. There was set times you could go and visit the doctor in Coniston but in these days when he was out 'on call', if anybody else was ill, you would get word to him and he would visit. Everybody pretty well knew one another's business then. The doctor would sometimes maybe get 'paid in kind' by receiving a bit of butter or eggs.

The road to High Park was at the edge of a wood off the Ambleside and Coniston road. At the corner of the wood where the roads joined, there was a story of a ghost being there. I passed there many times and never saw one. Once one of my uncles was walking home in the early hours and he heard a chain rattling behind him. He thought it was no good running, as it would go faster than he could. He carried on at the same pace and a dog came up to him dragging a chain, he was very relieved. The ghost was called 'Oxenfell Dobby'. I never met anyone who said they had seen it!

I followed the foxhounds when they were hunting in the valley. There was Willie Porter of the Eskdale and Ennerdale Pack and George Chapman with his Coniston Pack. Happy days! I went hunting on Wetherlam one Good Friday, saw two white hares, lovely animals, black legs and ears and snow white bodies. The cold winters of the early forties, killed them off.

I left school at fourteen and worked for Uncle Tom. Granddad had retired by now. We had no trouble in winter with frozen pipes, as we didn't have any. We got water from a well in the field about fifty or sixty yards away. The cattle shared it with us, it ran in an open stream and the cows walked in it. In summer if it ran dry, we went to another stream quarter of a mile away and if this ran dry, we had to go half a mile for drinking water. I can recall Langdale Tarn being frozen and people skating on it. We don't get many winters like those. I've walked the length of Langdale beck on ice.

Uncle Tom at High Park Farm, had cattle and sheep, not many of each. He milked seven or eight cows and took in visitors. We still lived in one of the two cottages which backed onto the farm. Uncle Tom sold milk to houses around. I carried it in cans, about two gallons in the morning and about the same at night, about two miles each way. I think it was done more as a favour than profit.

The Landlord was Mrs Heelis, (Beatrix Potter). I remember she came round in a horse and trap, driven by one of her men and visited the tenants perhaps once or twice a year and had a walk round. I can't remember much about her, other than she wore old clothes but she never bothered with us kids. We paid Granddad our rent and he would pay the rent at Mr Heelis' Office, he was a solicitor.

There was no haytime machinery, just a mowing machine drawn by one horse. We cut a lot of bankings with the scythe, it was hard work until you learned to sharpen it right, this was called 'wetting up'. We used 'wooden strickles' or 'whetstones' to do this. These were made out of a thin rail of wood, with a wooden handle, so you could get hold of it. You covered the rail with warm pitch and sprinkled on very fine sharp sand, which you found in some of the tarn bottoms on the fell. When I learned to use a scythe, it has a long wooden pole and when placed on end, it was about twice my height and the blade was four foot long. Some of those strong men who mowed with these big scythes, could cut a swath eight feet wide and two feet forward at every stroke. We used to get up early to mow while it was cool. The horses would be lying tired from the day before and would continue lying where they were, even when you were putting their halters on.

Some places in haytime would give you oatmeal and water to drink. This would be two handfuls of oatmeal added to the drinking water and brought out to the hayfield in a can, which had a lid and handle and was left in a shady spot to keep cool. This was very refreshing, we did not have all the fancy drinks we have now. Sometimes when we were tired in the evenings about six or seven o'clock, you maybe got a cup of rum and milk as a pick-me-up.

In haytime once one neighbour finished his hay he would help another. They used mowing machines made at Fell's Works, Troutbeck Bridge, near Windermere. These mowers were made for either one or two horses. It was quicker with a double machine than a single one. Most people just kept one horse as they took too much feeding to keep all the year round. Some farmers exchanged or borrowed a neighbour's horse to help to mow or plough. It made it faster and easier for the horses. We carted the hay on carts or on a sledge if the ground was steep. Some people made pikes of hay in the fields, about a load in each. This was alright in dull weather if the hay was not fit to go into the barn, it would be left in the pike for a while to sweat out (dry out in the sun). If the hay was brought in, in poor condition, it would sweat after being packed in the hayloft or shed, come out mouldy or could overheat and catch fire. Livestock will not eat mouldy hay. Some places piked all the hay before they started to lead any in.

I stayed one year with Uncle Tom, in this time granddad died, he was a good man, he did not smoke or drink. He had his different moods, he was either 'as daft as a brush' or 'as twinned as a cork screw' (a Westmorland saying for 'irritable').

About this time, I got my first bike, it cost a pound, from Charlie Raven (Uncle Tom's brother-in-law), at Elterwater. It was a bit big for me, so I took the seat off and wrapped a bag round the cross bar.

A school photo of me and Molly, when I started school.

CHAPTER TWO

WORKING ROUND AMBLESIDE

I left Uncle Tom and went to work for John Bell, at Round Hill Farm, a mile up Kirkstone Pass from Ambleside. I got eight pounds plus board and lodgings for six months. The food was good and there were three other workers, a man, a lad and a maid. Katie Hodgson, was the maid and she came from Dearham in West Cumberland, she was a lovely lass!

Mr Bell had sheep and cattle. He also had a milk round in Ambleside, which he delivered twice a day to. He had a contract with the Council for emptying dust bins, ash pits and earth closets, in Ambleside. Sometimes we worked on Kirkstone Pass carting gravel. We also helped the road men dig for gravel with pick and spade. In those days there were small quarries on the roadside, where you could dig for roadstone. We also could get sammel this way. This was like heavy clay soil with stones intermixed and would set like cement. There was a good sammel pit near Round Hill. The farmers would get paid by the Council for the removal of stones and sammel from their land. The sammel pit at Round Hill was fenced off separate and Mr Bell got a shilling for each load that went out (each wagon load usually contained one ton, cart loads were probably less), this was to cover for inconvenience and for the upkeep of the road. The roads then of course were not tarred as they are today.

Round Hill was a large house with six bedrooms, four stories high. They took in visitors. When the rooms were full, us lads slept in the granary. In these days the granary was used for storing farming tools, like spades, picks and hammers. One of our jobs we had to do, was ash pit emptying, which was a week's work for one man and a cart. If three went, we could finish by Tuesday dinner time, each one had his round. I had twelve earth closets to empty on every other Monday morning. These were all in a row and I would carry the contents in a swill, thirty yards to the cart. There were lots of those places, some had dust bins, some oil drums and some streets had a large pit which they all shared. We had to get inside and shovel it out. No one would do it today!

The contents from the dustbins, were carted with horse and cart to the Sewage Works, near Rothay Bridge, Ambleside. It was tipped in a field which was separate from the household treatment of sewage, to about ten feet deep, with a slope made by the horses and carts. The horses also usefully trampled it down. The roadside verges were wide then and the roadmen used to trim the edges and pile the dead leaves into heaps and we carted this on Fridays onto the top of the tip, so it was layered.

Of course, Ambleside was smaller then, in 1931. We got tips for taking extra things, like old bedsteads, furniture and such like. It was best about Christmas, when we got mince pies, cakes and other things to eat. Some hotels would give us a drink. Two of us were under seventeen at the time and the bins always seemed far lighter to lift after that! We were never caught drunk in charge! We used to have fun teasing the maids at the Guest Houses and big privately owned houses, and many at this time met their wives this way. West Cumberland and Barrow girls used to come and work in the summer months and some of course stayed the whole year round. They used to work in the bakehouses, cafes and hotels, looking after the visitors and many, as I said, married and stayed in the area.

The swill baskets were made locally from the coppice woods. These baskets were made from hazel and oak. The outer structure was made and strips of oak were woven in and out. Birch, sycamore and hazel, were grown in these coppice woods and would be harvested roughly every fourteen years. Birch twigs were made into brush heads and the thicker pieces into bobbins. Besoms were very popular with gardeners to brush leaves etc. These were brushes that had a large quantity of long twigs, roughly eighteen inches long, tightly bound at the top and attached to a long stick. These besoms were also used at the iron works at Barrow, when pig iron was made. When the pig iron was cooling, the scale (impurities) that floated on the surface, would be brushed off and because of the heat, besons only lasted the once. The sycamore from these coppices were used mainly for firewood. Hazel was used for pea and bean sticks and making yard sticks, these would be an inch to an inch an a half thick and used to stoke ovens. The items made from these coppice woods would be moved by rail from Coniston or Windermere.

Sometimes we went to the pictures at Ambleside, if we sat at the front, it was three pence. Talking pictures were just starting. We walked down the road singing, I found I had a good voice, it had just broken, happy as sandpipers we were! When we returned, we walked up with the lamp lighter, George Black. There was a street lamp half a mile up Kirkstone road, it was gas, for Ambleside had its own Gas Works which were situated at Blue Hill, above Fisherbeck. The lamp lighter carried a pole with a hook on, to pull the chain. Sometimes we stayed and had a talk with him and he would smoke his pipe. When the Church clock struck eleven, he started off on his round, he had a long way to go.

Once there was snow and ice on the road, Mr Bell set off with the milk, the horse slipped on some ice, broke both shafts off the float and finished on its back in a ditch. We had to pull the wall down and pull it out with another horse, she was none the worse, only lost a bit of hair!

I stayed there two years, then I moved across the valley to Mr Parsons at Grove Farm, a much bigger farm, a thousand sheep and milked forty to fifty cows. There were five sons, one was still at school, the two younger sons spent most of their time on the milk round. We got up at five o'clock and worked till six thirty or seven. I got eighteen pounds for six months. The boss and two sons started off at seven o'clock with milk and arrived back about ten thirty. In the evening they started at five o'clock and back about seven. They had three churns (or milk kits as they were also known) in the float with taps on, they filled four gallon containers and measured it out with a pint measure into people's jugs, there was no milk bottles then! Some days there would be some milk brought back in the kits, maybe not as much was required at the hotels. People did not have regular set orders and the amount required varied from day to day. The milk that was brought back was not wasted. The milk was separated and the blue or skimmed milk was given to the calves and we had the thick creamy milk. The cream did not often have time to settle on the top of the milk. I remember one of our treats we had as a result of this, was strawberry jam and cream, eaten along with bread and butter. We ate well!

Delivering milk was an awful job in bad weather. In those days the price of milk was two pence in summer and two and half pence in winter. The half pence difference was to cover the extra cost of hay and feedstuffs. It made seven pounds difference each week, which was a lot then.

Mrs Parsons was a grand woman, she had nine men in all to feed, husband, five sons and three workmen. She worked hard, washing and baking for all these people and meals were never late.

Mr Parsons was often away, he would give you enough work for two days and leave you something to do if you had any spare time. He went to Kendal to buy cows when the milk supply was low (various reasons for this, cows giving less milk, due to calve, more calves or more milk required by customers), he always bought prize cattle, paying about twenty-eight pounds each. There were three horses and a milk pony. The pony was fitted with studs on its shoes to stop it slipping on the ice. I was the horseman and I was only seventeen.

We had meadow land about two miles away. I carted muck there with two horses and carts, the road went down one side of the valley and up the other. It was steep so I used a trace horse. I drove one horse, lead the other and the trace horse tied to the last cart. When I got to the bottom of the hill I tied one horse up to the fence, yoked the trace horse to the other cart and up the hill we went. When I got to the top, I tied the cart horse up and took the trace horse back to the bottom and repeated the action again. When I reached the field, I tied the trace horse up, put the others seven strides apart, pulled one heap out of each, led one horse on seven strides and told the other to follow. It stopped when I stopped, keeping a straight line. I had all the heaps to spread later.

I used horse language. A-ah for left and gee-back for right. These trace horses were used as extra pulling power (horse power). They were actually attached by lengths of chain to large hooks on the shafts of the cart that the original horse was pulling, it then walked a short distance in front of this horse and just acted as extra pulling power for heavy loads or going up hills. Instead of having four horses to pull two carts, you had three and kept on transferring the trace horse when extra pulling power was required for going up hills etc. I only got two trips in a day as I helped with cattle night and morning and there were three field houses to go to twice a day.

At nights when it was dark we had paraffin lamps, they often blew out in the wind. We mixed meal twice a week for the milk cows, Mr Parsons had his own mixture, it was done last job at night. We had to wash our boots well or the rats would eat the laces when we were in our beds! There were rats running about in the house when we got up and we often had a rat hunt with the fire irons (fireside poker and tongs). We only went out at weekends and we went in our work boots to Round Hill because the road to there was muddy, we changed and left our boots behind the wall. Haytiming was still done by hand, sledging and carting, but we had a much stronger team than up Langdale. Once in haytime we were working in a field joining the yard, when dinner time came, someone on Kirkstone road timed us. We were only out of the field eight minutes and dinner was ready in our plates when we went in. I knew it was a hard place before I went there, but it was a challenge. I only stayed a year, it was plenty!

North Road, Ambleside in the early 1900's - I went up this road often.
Note the gas lamp on the right.

John Marsh Collection.

CHAPTER THREE

MY YEARS AT SELSIDE AND RAVENSTONEDALE

I came to Kendal for the first time in my life in 1934, it was 'Hiring Day'. The hiring was done between the top of Finkle Street and Market Place, on both sides of the street. There would be dozens of lads and lassies. There would be two Hiring Days a year, Whit Week in May (Whit Saturday) and Martinmass (11 November). You would be hired for six months and then you would go for hire again, unless of course your boss wanted to keep you on and you had struck a bargain.

I only knew one person there in 1934 and that was Stan Wilson, from Ambleside. Stan introduced me to Mr Blenkharn. He had worked for him a few years before and he was doing us both a favour. I was a good worker, Mr Blenkharn had a good reputation as being good to work for and had a good place. Stan had gone out of farm work and was just there to see people he knew to talk to.

I don't know the name of the lad, but a story of one of these Hiring Day's goes, that this lad was talking to a man who wished to hire him. He asked him if he had a reference and he said, "No," but he would go and get one off the farmer he'd worked for. He came back to the man, who said, "Have you got your reference?" The lad said, "No, but I've got yours and I'm not coming!" You used to get a shilling if you were hired by that person and that was called 'binding money'.

I was hired by Mr Blenkharn of South Gateside, Selside. It was November, very foggy weather, the A6 road was three hundred yards away and I never saw it for a fortnight. He milked twelve cows, made butter and reared calves. The milk went through the separator which was turned by hand. Then there was churning day and I did that. The butter was sold in Kendal Market on a Saturday. He farmed Hause Foot Farm, which lies at the bottom of Shap Fell. He had a man in there and it was all sheep. My first pal I made there was Jack Tallon, he was about school leaving age and we are still good friends.

There was ploughing to be done which I was not used to. I soon got the hang of it. Ley corn was first, then it was left fallow in winter (fallow means the land was left to rest), then worked (made ready by ploughing etc.) and sown with turnips to feed the cows and lambs. Cob lime was mixed with soil on the headrigg (this is land not ploughed around the edge of the field), this was called lime and mould. This lime and mould was best mixed during the winter. Sometimes we would have a large pile of lime in which soil would be mixed, also grass sods and trimmings off the roadside verges would be added to increase the volume. With the chemical reaction from the lime this would work during the winter and when we had time we would occasionally stir it up to keep the process going and this would be put on the land after harvest in the autumn. The cob lime was the rough lime which was burnt in lime kilns and we got this from Penningtons Quarry, in Kendal and it came in five ton lorry loads. This we loaded into our own carts and put out in heaps. It was called 'slecked' a Westmorland term if it got wet. We tried to spread the lime on before it got wet. It was terrible stuff for burning

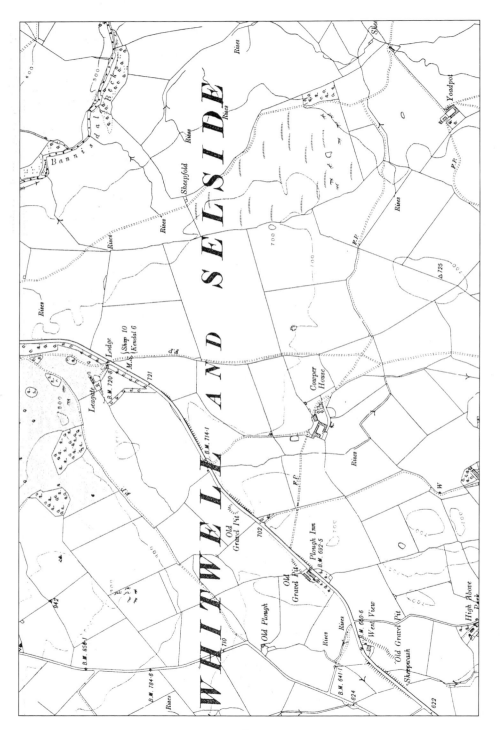

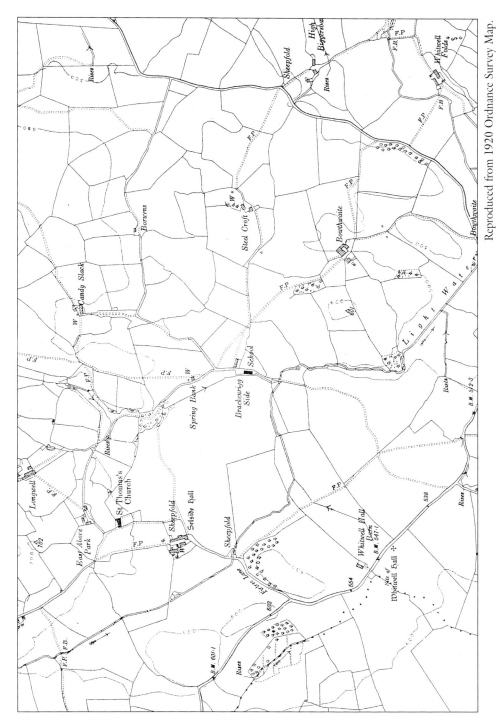

Reproduced from 1920 Ordnance Survey Map.

29

your hands or where the sleeve of your jacket rubbed your wrist. Kibbled lime which you also hear of, is a finer more refined lime.

The field was sown with corn again and undersown with grass seed. When the corn was taken off, the lime and mould was spread on these seed plants. With it being under-sown, there was no need to plough again, as the field was now covered in grass. The following year it was spread with basic slag (this was darker coloured than lime and was the dust from the steel works), this and the manure from the lambs, meant you had a good field for ten years or more. This is called crop rotation.

When we cut the turnips in the back end, we brought them in and would 'hob' them. This was when you would make a big pit, by putting them in and covering them with straw and soil on top, to protect them from the frost. Some rows of turnips were left intact out in the field and you would give them to the sheep, so many rows at a time. The turnips were also given to the cattle by chopping them. We had a turnip cutter. There was two types. One with just a large blade that came down on top of the turnip when you pulled the handle and cut pieces off. Another was when you placed them into a container, that had a large drum barrel, that turned when you turned the handle and pieces of turnip would drop out underneath into a container.

When we began to cut corn, we used a horse drawn mower, it had two seats, one man drove the horses and the other raked the sheaves off. There was a rack on the cutter bar to catch the straws and he had to rake this also to stop the blades from getting jammed. When there was enough on the rack for a sheaf, he would press a lever with his foot and rake off the straw in a neat pile. There was two or three people who walked alongside tying the sheaves, they got a few straws, put them round the bundle, twisted the ends together and tucked them in. The band was not very nice if there were thistles in! The corn sheaves were stooked on their ends, in two's, six or eight to a stook. This was led in and either put in the barn or stacked in rucks in a stackyard.

The steam thresher, Houghtons from Burton, came to thresh the corn in the autumn, usually October. The corn sheaves which were in the barn lofts, were thrown down and into the thresher. The sacks of corn and straw were stored away for later use. The steam thresher continued round the Parish and everyone helped one another. All the hired men and lads followed round each day, it was about a week's work. We had a great time leg pulling and arguing, seeing who could carry the biggest forkful of straw. There was always a good feed on threshing day, I think the farmers' wives tried to beat each other with the best display of food. We didn't complain, it was lovely!

We had turnips in a field below Watchgate, they were a very good crop. I cut one once that weighed seventeen pounds and others weighing between eight to twelve pounds, which was still pretty heavy. When I had been at Selside a month or two, I got pally with other lads. Most farms had a lad, some two and some also had maids.

Grandmother Bowness died about this time and it was her wish to have four black horses pull her hearse, this would be from High Park, Little Langdale, to Langdale Church about three miles. She had spent all her life with horses and wished to end it this way. Her wish was granted, although there were of course motorised ones by now.

When I was hired by Mr Blenkharn, I was paid at the end of six months and if I needed any clothes during this time, I went to Blands, in the Shambles, in Kendal, and if you told them who you were working for, they would give you credit. This was one way of doing it, another was that your boss settled the bill and knocked it of your wages before you got paid.

My best pal was Tommy Mooney, he was a good singer, we joined the choir at Selside Church, the attendance was very good. There was only the Church and the Plough Inn to go to, and the Church was cheaper! Tommy and I biked to Kendal and to dances all around the district. When we went to Kendal, we would leave our bikes in Parkinson's Yard, Kent Street. We would come back for them about half past ten, there would be thirty to forty bikes there, not a lamp or pump missing! There was a cottage at Stocks Hill, Sally Whitwell and her sister lived there and sold pop, fags and toffee. The lads would call for a rest when biking up to Selside, they liked us to call and quizzed us about the Parish. Some dances we went to didn't finish until half past two in the morning and by the time you cycled home it wasn't worth going to bed!

Tommy was one of the two that survived the bomb at Cooper House Farm, Selside, in April 1941. There was thirteen in the farmhouse that night, the family consisting of Mr Wood, his mother, wife and two children, Tommy and another worker and six evacuees. The bomb fell directly on top of the farmhouse and all were killed, except Tommy and his pal from Staveley, who are both now unfortunately dead. They were both flung by the force of the blast up into a field on the mattress they had been sleeping on, with the bed-clothes still on. Tommy had a broken leg and the other lad was more shocked than anything. Both recovered to continue their working lives. There was a big monkey puzzle tree in front of the house that was snapped right off. There were a few slates off the byre and barn but no stock were hurt. The farm house was completely flat with the walls containing and taking the full force of the explosion. Nobody knows why the bomber dropped the two bombs, one apparently dropped elsewhere on the fells. There would be no lights on, the only thing being that perhaps the pilot knew he was in the area of the Haweswater Reservoir, or the main railway line or the important main A6 road.

When we were in Church and sat in the choir stalls we could look down the Church and see who was there. One bonnie lass came every Sunday and I thought to myself, "She will do for me!" Then when I met her at Selside Hunt Ball, we danced all night and I walked her home to Bowthwaite Farm, where she worked. It was not serious then, we were both shy people and we met only now and again.

One Christmas, Tommy and I went to the Plough Inn, had a drink or two and we set off home. Tommy was going to Whitwell Fold and I was going to Gateside. I walked with him to Above Park and then crossed the field. There was a stream to cross with just a slab for a bridge and it would not keep still! It either went from side to side, or up and down. In the end I went across on my hands and knees! I had been at Gateside two years before I left and went to work for Mrs Dargue, at Forest Hall, for six months. I enjoyed it, there were thousands of acres of land to walk and lots of sheep.

Hiring Day in Stricklandgate, Kendal (Finkle Street to Market Place) in early 1900's.
There weren't as many people there in 1934.

Museum of Lakeland Life and Industry, Abbot Hall, Kendal.

Cooper House Farm, Selside, April 1941, after the bomb dropped on it, killing Mr James
Wood, his family and six evacuees - there were two survivors.

John Marsh Collection.

Photo of farmer's bikes in Redmayne's Yard, Stricklandgate, Kendal in 1930's, where farmers lads also used to safely leave their bikes to enjoy a night out in the town - this is similar to where I put my bike in Parkinson's Yard, up Kent Street - but this is a much larger yard.

Margaret and Percy Duff Collection.

Inside Selside Church where I first saw Maggie and sang in the choir.

John Marsh Collection.

I then went to work at Gateside, again for six months. I started to walk my old girl-friend out again, Maggie (Margaret Isobel Potter) she was called. I was surprised she was still interested in me, I guess I must have been attractive!

In May, I went to work at Bainsbank Farm, in Middleton, near Kirkby Lonsdale, as a shepherd, a very good farm on the banks of the River Lune. I had five different work mates in six months. I was cowman and horseman sometimes. Tom Robinson was the farmer I worked for. The Lunesdale Foxhounds were formed that summer and kennelled there. He always asked me to go back to work for him every time he saw me.

When I worked at Bainsbank the stallion and groom travelled round the district visiting farms. He stayed the night at Bainsbank and mares came to visit him. I heard a story when I was young, about a groom and horse. The story goes, that the horse only sired one foal but the groom sired two children.

One Sunday morning at Bainsbank, I went to the allotments at Lupton to look over the sheep. We'd been haytiming and had not had time to go before. As I left the farm I met a few cars going up the lane. When I got home in the afternoon, I went to the back barn for something, there I found a ring of bales and many feathers. We lived off game cocks all that week!

Maggie still worked for Mr and Mrs Sutton and went with them when they moved to Above Park, Selside, to farm. I biked to see her on Sunday nights, there were no Sundays off. Our friendship was getting serious. It was sixteen miles each way to bike. The first time I biked there, I got back about midnight, my supper had been left on the table and the terrier dog had eaten it. I got my supper at Above Park after that. It was too far to bike and winter was coming on, so I moved back to Selside, to Bowthwaite, and worked for John Armistead. It was a very good place to work and I went to see Maggie twice a week.

Mr Sutton, at Above Park, was a cattle dealer, I often helped him to castrate bulls and I got so involved I went to work with him, it was an excuse to be near Maggie. This was three changes in eighteen months. There was a saying, men who only stayed six months were no good, but you learned different ways.

I learned to drive in the first few months at Above Park. I passed my driving test second time, testing was just starting then. I went to Preston with the car and trailer for calves, which were later sold round the district. I have walked cattle and sheep from Selside to Kendal Auction many times, both ways. There was less traffic then but some vehicles had very poor brakes. One firm from Scotland, took fish to Manchester and Liverpool. The wagons came down about midnight and travelled back up at midday. They changed drivers at Kendal. You had to look out when they were on the move.

When I had been at Above Park eighteen months, Maggie went home to Brockstones, Kentmere, to help her mother. There were eleven in her family, the older girls took it in turns to go home to help. I went to see her once a week, twice if I could borrow one of Mr Sutton's cars. Other times I biked up Longsleddale and walked over the fell, it

Sheep shearing Rough Fell ewes at Bannisdale Head in the late 1930's. From left to right, an unknown hired lad working at Bannisdale/Dry Howe; David Smith, gamekeeper at Thorn Cottage; Billy Hawes (brother to Kit, my friend) and a shepherd at Bannisdale Head; Tom Dixon, shepherd at Forest Hall and George Davies, shepherd at Dry Howe.

Photograph by Joseph Hardman.

Supplied by Cumbria Heritage Services, Kendal Library.

Our wedding day - it was a lovely summer's day. The photograph was taken outside Town Head, Long Marton.

was a bit lonesome walking over Stile End road about midnight. The family moved to Town Head Farm at Long Marton, Appleby, in March 1939. Maggie went with them and it became too far to go courting, so we decided to get married. We were married on 1 June 1939, at Long Marton Church and Alf my cousin was best man, Elsie (Maggie's sister) and Dorothy (my sister), were bridesmaids. It was a quiet wedding, just our families, no big fuss but the marriage has lasted fifty-eight years. Mr Sutton loaned us one of his cars. We had four days in Scotland. Not many people had a car for their honeymoon in 1939. We stayed at Above Park till October and I got a job with a house. We wanted to be on our own as Maggie was pregnant.

We went to Ravenstonedale and I became shepherd to Mr Handley, a big farm at Park House. We had a cottage in the village, the rent was three shillings and six pence a week. My wage at the time was forty-five shillings a week, milk and potatoes thrown in. This wage went up yearly and by the time I left three years later, I was getting £3. 10s. 0d. This was a large farm with over a thousand acres and fell rights. I had a sheepdog called 'Spot' that I gave ten shillings for and a pup, that I'd been given, that I trained. We lambed six hundred sheep and milked fifty cows. We had over one hundred cattle altogether. They kept a lot of hens as well and sold hatching eggs.

I had three workmates and started work at six in the morning. I biked home for breakfast and dinner and took a packed tea. We took some (300) sheep to the fell for a few weeks after the New Year. It came a very heavy snow fall on 16 January 1940, Mr Handley and I went for them the following day, two miles up the fell. We found them all but twelve. I went to look for these a mile further on the fell, I found them altogether. I started off home, three miles the going was slow, it was dark before I got them to the farm. There was a field house a mile from the farm, I had to care for the cattle there. These field houses usually housed cattle and hay to feed them with. The cattle had to be regularly fed, watered and cleaned out. One of my mates had been feeding and looking after them. The building was called Wuthering Howe (Wuthering Heights) but I never saw any Kathy!

I walked up the village tired and hungry and got home to find the nurse and neighbour were there, the baby was on its way. The doctor managed to get through the snow and we had a baby girl by ten o'clock. When we were in Scotland we stayed at a farm one night, she gave me two duck eggs, I blamed them for the upset! The baby was very frail, she was two months premature. The vicar came and christened her, we called her, Mary Margaret. They rubbed her in olive oil and wrapped her in cotton wool and she was kept like this for three months. When she was a fortnight old, she got pneumonia, the doctor gave her some M and B tablets, they had cured Mr Winston Churchill. These pills were new out, she had to have a quarter of one, twice a day. She was only three and a half pounds in weight when she was born and then went down to two and a half. She got over that and was well above average at twelve months. Coal was rationed and the snow lasted six weeks. It was a problem keeping the baby warm and then the water froze up. I dragged milk kits (churns) of water from the farm on a sledge.

In June 1940 we got an evacuee, he was from Newcastle, Ronnie Eltringham was his name. He still keeps in touch and calls on us sometimes.

The winters were bad when I was at Ravonstondale, snow and frost for weeks. There were hundreds of rabbits, most of them perished in the snow. 'Spot' my dog could find a sheep six foot deep in a drift. We had some outlying cattle in 1941 and one morning they were behind the wall with just their horns showing. They soon broke out when I shouted. The sheep were foddered (fed) well but grass was very scarce in the spring, some had a lot of work with the lambs. I was among them by first light and often suckling and feeding in the barn when the Church clock struck midnight. My mates had gone home at six o'clock. I was twenty-four then, I was strong and fit.

Spring brought seed sowing and I helped with those jobs when lambing finished. Then it was turnip thinning and we had seven acres, including mangolds. We thinned the turnips by hoeing with a hoe. When I was at Selside we thinned the turnips out by going down on our hands and knees and thinning by hand. It was called 'creeping'. We folded a hessian sack, then wrapped and tied them round our legs to protect our knees.

Clipping time was next, then haytime and harvest, long hours and no overtime. One summer, I think it was June 1941, there was a plague of caterpillars on the fell, they ate all before them, but did not go on the wet land. I remember they were dark brown in colour, with a yellow or green coloured circle on them. We had them for a few weeks. One day I took some tea and a flask with me but when I sat to eat it, I had to close the lunch box or it would soon have been full. The seagulls were sitting in the sun, their crops filled with caterpillars. Shallow parts at the edge of small streams were full of them floating dead. When we went to clip the sheep, the fleeces were full of dead caterpillars. I don't know what they turned into, but there was a lot of granny long legs around that year.

I attended a Shepherds Meet at this time, which was held at the Temperance Hotel at Ravonstonedale, as Mr Handley was away. A Shepherds Meet was where farmers or shepherds in that valley or dale, used to take the stray sheep to in the autumn and exchange their strays. They would have these Meets at well known places, the most noted being Mardale, Thirlmere and Loweswater. Every year you could well have a few strays if you had heath sheep on open heath land (where there were no fences). Shepherds would be able to identify their own sheep by their own different markings. This was of course before people had tractors, trailers and cars. Mr Handley I remember paid a small subscription, this was to cover any costs and on the day after the business of exchange was over, the shepherds and farmers would retire to the pub, for sandwiches or pies, and a drink and a sing song. At Ravonstondale with it being a Methodist area there was no alcohol but we still enjoyed ourselves afterwards.

I joined the Church Choir. The Home Guard was formed. We all had to join, there would be twenty of us. In summer two of us went down Smardale Gill to guard the viaduct, we had one rifle and five rounds of ammunition, might as well have had a water pistol! We took it in turns, it came around about once a week. We did drilling and target practice on a Sunday, we had a range made up on the fell. We had four men that were Officers, they had been soldiers in the First World War. The other duties were patrolling the village. We also used to practise guarding the village, where some would be patrolling and others would try to get in without being seen.

One of my work mates was called Jim Pratt, he was a good sport, we went catching rabbits and grappling fish in the fell becks. My old dog liked to go rabbiting, she never looked at one when she was working, but if I looked in the walls she would set them (she would smell the rabbit, stop and look, putting her head to the side), I got many a dinner. Meat was rationed, it helped out with a bit of rabbit. I had nothing to do with the black market but I knew it went on.

It was much better when I had a home of my own and a wife to look after me. It was a bit rough in farm service. We'd only one lot of work clothes and we often got wet and laid them out on the bed to dry while we slept. They were damp and cold to put on in the next morning. Life was very pleasant at Ravonstonedale, the people were kind and friendly, but I still thought of Selside.

An early picture showing sheep shearing at Mr Robinson's, Bainsbank Farm, Middleton. With upwards of thirty clippers, 800 sheep would be clipped in one day.

Bannisdale Head when we lived there. "We must have had visitors, the sitting room fire is on."

Bannisdale Head with some Rough Fell sheep and me standing in the background.

CHAPTER FOUR

BANNISDALE HEAD AND BACK TO SELSIDE

When Bannisdale Head changed tenants, Mr Metcalfe took over. He wanted a shepherd, so I applied and got the job. Ronnie went home to Newcastle. We moved to Bannisdale in October 1943, it was a change to living in a village. Our nearest neighbour was two and a half miles away. There was no electricity, so we left our wireless and sold it to Maggie's parents. We were six months without one. It was back to candles and paraffin lamps. We went shopping to Kendal by bus. We went four miles by horse and cart and left it at Above Park until we returned. We did this for one year and then Mr Metcalfe got an Austin car, he built a cart on the back half and we just went to the main road, but it was much better than the horse.

There was five hundred lambing sheep and about twenty cattle more in summer. The first winter we had at Bannisdale was very mild, frogs were spawning in the spring in the yard and it was only 14 February, it was never frozen. It was a very warm lambing time. I remember there was six inches of snow on 12 May but it thawed all in a day and made a flood. The wood at Bannisdale was felled the first year we were there, two men stayed with us and three horses were stabled. The wagon which took the trees away widened the road and I had to infill it with stones. If I had any spare time I went repairing the road, it was two miles long and it was only tarred from the road at Dry Howe to the A6.

Bannisdale was the best job I ever had before farming on my own. Mr Metcalfe was a good boss, I did as I liked, he trusted me. The sheep were Rough Fell, a docile breed of sheep, tough and strong. We let the tups go out to the ewes on 9 November, we reckoned four days back and five months forward, that was when they would be due to lamb. Three hundred had Rough lambs and two hundred had halfbred lambs, got by a Teeswater Ram. The lambs were called Mashams. The Mashams were sold at Kendal Auction Mart in September, most of the gimmers (females) went to breed fat lambs in the lower districts. The wethers went to feed, these are castrated males. The male Rough lambs went to the lower farm, Biglands at Patton, where Mr Metcalfe lived. They then went for slaughter when they were fat.

The gimmer lambs went to winter on farms around Kendal, in numbers of about forty or fifty. I think the price for wintering was ten shillings a head, as the sheep prices rose so did the cost of wintering. These lambs were for replacements to the flock. Female lambs were known as hogs in their first year. After they were clipped they were called shearlings and had their first lambs at two years of age. After this they were just known as ewes. They usually had a Rough lamb for another two years, this was got by the Rough tup, then a halfbred (crossed with a Teeswater tup) for one year and then sold as drafts or cast ewes. Sometimes if a ewe was still in good condition, she would be kept on to lamb for another year. Ewes were selling at auction from £2.10s. to £4, halfbred lambs would make up to £4, with some of the best making £5.

Picture taken of us all Whitsuntide 1947, at Long Marton, me with Brian, Mary in the middle, Maggie and Dorothy.

Me, Brian, my two sheep dogs, Kep and Ken in the sheepfold at Bannisdale Head.

We dipped in the autumn, spring and August as a fly repellent. Dipping in March cleaned the sheep for lambing time and if they were handled right they came to no harm. Sheep don't get dipped enough these days. I have been among sheep all my life and I've never seen a sheep with scab. I've been lucky to work where the sheep were dipped right. We used Boon's Fly Dip and Border Paste. The Border Paste was used in the autumn for sheep tick and scab. We speaned the lambs (weaned - separated the lambs from their mothers) after the 12 August (The Glorious 12th!), so as not to inter-fere with the shooting, as the gathering in of the sheep and lambs would disturb the birds.

While I was at Bannisdale I helped at Forest Hall and Dry Howe on clipping and gelding days (castrating). I helped to gather at High House, this was part of Forest Hall Farm and both these farms formed part of the Levens Hall Estate. Bannisdale and Dry Howe, were owned by Fothergills and were part of the Lower Bridge Estate. I met the shepherds at five o'clock in the morning on the fell top and we finished in one day, but it was a long one. We gelded the lambs, cleaned the ewes and attended to any feet that needed it.

I recall one day it was very hot, we had finished in the evening when it started to thunder and rain very heavy. We did not have a jacket and there was no shelter. The hogs were all sorted from the ewes and lambs. They were all in a long drove beside the wire fence, five hundred of them. There came a loud crack of thunder and every sheep rose a good yard off the fence in one move. The lightening struck the fence and we saw it travel along the wire. There were no sheep hurt, they were clipped next day but I noticed some of the fleeces were singed.

I went on clipping days to help our neighbours and they in turn would come back and help us. We had about seven or eight clippers at Bannisdale and would do about five hundred sheep in one day. They were made up of relations and friends. One person had the job of catching, another had to roll the fleeces once they were clipped and one to mark the ewe with the farmer's own markings, so he can identify his sheep from anyone else's. This mark would be a certain colour on that particular part of the sheep, usually called a pop mark. At Bannisdale a black ring was put on after the sheep was clipped, by a metal stick. The black pop mark was put on the far shoulder or near hock (in autumn). The ears had a punch hole (hole made in actual ear by hand punch) and on their horns the farmer's initials, ours was 'BM' for 'Beadle Metcalfe'.

At Forest Hall there would be about thirty clippers, it was all done on stools. There would be two markers, two catchers and three fleecers (rolling the clipped fleeces into a tidy bundle). There would be two thousand sheep clipped by hand in one day. Bannisdale sheep grew very heavy fleeces. One year we had a geld sheep (a ewe that did not bear any lambs that particular year) and it weighed in at ten pounds.

The women folk all joined in to help with the meals. We got three meals on clipping days and all were very good. They used to hold gatherings, called 'Clipping Nights', when people sang and danced, but this was before my time at shepherding.

The first few winters were not too bad, we were snowed in at times but not for long. We got a monthly order brought up by Leightons Grocers, from Kendal. We stocked up in winter, bags of flour and sugar, oats etc. There was hen feed, dog feed and some for the pigs. We killed two pigs, the meat ration did not go far. If we were blocked in by the snow when the order was due, I would go to the Plough Inn with a horse and sledge.

We went to Kendal each Saturday for yeast and odd things we needed. I walked if the snow was on. We got our wireless battery charged at Selside Garage and exchanged it every fortnight. We decided to have more family in 1943. We had to try and plan it for summer, when there would be no snow. It was due in September, not bad planning, must have been the Christmas spirit!

In June 1944, Maggie was ill, she had a threatened miscarriage and was in hospital for a week. The doctor would not allow her to go back to Bannisdale, he said it was too far out and we had no telephone. Maggie and Mary stayed at Above Park with the Sutton family until September and I was on my own for three months. I remember I went a whole week once with only the dogs to speak to. Dad and mother came for odd weeks, so I had a lot of time to think. I wrote my first poem then and called it appropriately, 'Solitude'. I've written twenty more since then. When you are on your own with nature, you get the inspiration, it's a form of talking to yourself. I was always whistling or humming some tune and gradually started putting words to it. I have won trophies at Merry Neets and Gatherings with my songs.

Dorothy was delivered by caesarean section and Maggie was very ill in hospital for three weeks. I went fishing one evening when I was on my own, I caught ten nice trout, cooked them all and ate them for my supper. When they arrived home, I bought a car for £60, a Hillman Minx, I was getting four pounds a week wage now and it was better for the baby. When I was first taken on at Bannisdale, I got £3 a week, potatoes, milk and the house.

We had postal deliveries three days a week, sometimes we would get a funeral card and the funeral had been. Mary started school in 1945 and I took her the three miles to the A6 to catch the school taxi and I got paid for doing this, but in summer there were eight gates to open and shut.

In 1946, Brian was born without any planning. It's surprising what you can do when you're not trying and we were pleased to have a son but it was another caesarean birth and all went well. The winter of 1946-47 was the worst in living memory, following a very wet summer. Fodder (feeding stuff) was scarce everywhere and bad to move on the snowy roads, I ran out of hay in February. The cows had nothing for two days, it was a worrying time. We got some hay delivered to Leagate Barn and took it up on the sledge, this happened a few times. All the hollows on the fell were full of snow, the hill ends were blown clear of snow and the heather was showing through.

I went up the fell nearly every day. I walked over the drifts and the sheep followed my footsteps and they soon cleared off any shoots that were showing. The car was not out

Dad and Mum at Bannisdale Head about 1950, they used to come for their holidays and Dad used to enjoy going fishing.

Photo taken for me by Joseph Hardman, who sometimes came to the farm to take photos. This is Mary feeding the hens and ducks outside the front door at Bannisdale Head, she was only four.

for nine weeks, it was wet, wild weather before the snow came. Maggie did not go with me to town for two weeks before the snow came because the two babies were small and she was never off the place for twelve weeks. Maggie never grumbled!

I walked for shopping in the snow for nine weeks to the main road and got the bus or a lift into Kendal and carried the shopping back in a rucksack and bags. I walked each week and I noticed there was a boggy place by the beck. I saw this black mark in the snow get bigger each week and this was two hundred yards from the road. One week I went to look at it, the snow was about thirty inches deep and I discovered it was a warm spring, not frozen and in it were dozens of frogs. The foxes had been catching them and splashing the peat on the snow.

Mary went to stay at Patton with Mr and Mrs Metcalfe, so she would not miss too much school. She had to walk three miles in snow to the car and she was away many weeks. The weather was wet and cold in lambing time and we lost forty ewes and had only one third of the normal crop of lambs. There was plenty of firewood after the woodcutters had been. Peat was the fuel in these fell places. It took a lot of drying and when the summers were wet, I would get wood.

One January, I remember the snow was lying on the ground, it was frosty but we had sunny days. One afternoon I heard this fox barking, I looked up and saw it trotting across the fell, it did this for three days. I wondered why it went the same time every day. This carried on for a fortnight and then I understood why. I saw it lying under this rock and as the shadow moved up the hill and the sun moved away, he also moved. I knew it was a 'him' by the bark. I could tell the time of day in sunny weather, just by looking at the sun moving up the hillside.

I had eleven lambing times at Bannisdale and never lost a lamb to the fox. The Pest Officer arranged for Drives and all the farmers joined in with beaters and guns. We would get five or six in one Drive at Bannisdale. The hounds did not catch many foxes, the feed they had was poor and they were not as fit as they should have been. I caught many foxes and cubs. My neighbour at Dry Howe was called Dick Packham, who was about my age. We had the same interests and went cubbing together. One spring I remember we caught thirty-eight. There was no game keepers then during the war, they were away fighting. Dick was like a brother, he was a great guy, poor chap died at fifty.

I had a good terrier and she went with me to the fell. She visited every hole and if there was something in it, she would give a few barks and I knew where she was. We had to control foxes somehow. There was a bounty on foxes' tails in war time, half a crown for cubs and ten shillings for an old fox.

The winter of 1953 was very cold, Brian and Mary were playing on the ice under the bridge at the bottom of the yard, when a hen flew onto the bridge wall and knocked a stone off onto Brian's head, fracturing his skull. The road was blocked with snow drifts, so I loaded bracken (cut and dried from the fellside and used for bedding cattle in winter), onto a sledge and Maggie and Brian sat among it. It was a good steady horse. I had to lift the gates off their hinges. Brian had also bitten his tongue and it was cut right across and bleeding a lot.

When we got to Dry Howe, Mrs Swift took them on the tractor to Selside Garage and from there to Hospital by taxi. Three days later he had to go to Manchester to have an operation. There was no phone, so the Ambulance men came to look for me at Bannisdale. It was dark and they called at the Plough Inn and spoke to Mr Dargue, who was there. He said, "Go to Forest Hall and my man will take you there in the jeep." They wanted my permission to move him and they arrived about seven o'clock. I went with them to Kendal Hospital and then to Manchester. They brought me back to Lowbridge, it was four in the morning, I walked four miles in the snow to get home, the worst night of my life. They took the damaged bone from his head, he was there three weeks and came home with a plaster on, covering a hole about the size of a half-penny. He went back in six weeks and they grafted a piece of bone from his hip onto his head and he has never been troubled since. This was the end of Bannisdale for us, we left in June that year. By the time I left I was making £7 a week.

I got a job at Grayrigg but could not get a house, so that fell through. We moved in with Mr and Mrs Hewitson, at Above Park, Selside, they were good friends. The Sutton family had moved to Northampton. Above Park seems to have been our salvation, I took casual work on farms and Maggie did domestic work. I started working at Kendal Auction for three shillings an hour. In the autumn when the sheep sales were on, we often worked until midnight and made ten or twelve pounds a week and thought we were well off. I got more work than I could do in summer but no one wanted you when the weather was bad. One winter when I was on casual work, the road was full of snow from Above Park to Spring's Gate. I dug it all out on my own, it took me nearly a week, hard work but it was a job and I could do nothing else. I went thinning turnips in summer and haytiming. I had haytime places, first at Carus Green, near Kendal, then Bowthwaite in Selside and Harewood in Whinfell, each one later than the others.

I also looked after the Church heating at Selside. There was no electricity then, just a coke burning stove under the Church. Hot air from this stove went into the pipes under the floor and up through the iron grids. It was not very good in cold windy weather. I got jackdaw nests (full of dry twigs) down out of the tower to light the stove. I went to light it on Saturdays about six o'clock at night, went again at ten o'clock to stoke up and again at five o'clock on Sunday morning. My pay was ten shillings a week for doing this.

Kit Hawes was my working partner at the Auction and together we went, clipping, walling and any other work we could get. He was a good workmate and always pulled his weight (did his share), often more. We often had a day off in winter and went hunting on Saturdays usually.

CHAPTER FIVE

BORROWDALE HEAD TO MY RETIREMENT

Dad died in 1954 and mother a year later, they were only seventy years old. In 1956, I rented and took over the farm known at Borrowdale Head, at the foot of Shap Fell and owned by the Bagot family of Levens Hall Estate. It is a hundred and twenty acres, mostly rough grazing, with twenty-five acres meadow land. I was forty years old, a bit late to start on my own. I had spent my best years working for other people. We moved in on the 12th of May. My capital was small, £350. I borrowed £500 off my father-in-law, and paid it back over two years.

Mr Dargue at Forest Hall, gave me two implements he had finished with and I repaired one with the other. He bought a new mower and I used it as well and then he gave it to me. I paid him back by helping to clip and other jobs. I got an old tractor off Mr Potter, Maggie's Dad and paid for it bit by bit, as I could afford it. It was a grey Fergie TE20 TVO (Tractor Vaporising Oil). It was called this as you used a little petrol to get it started and once it warmed up, you changed the mixture over to paraffin. I did not borrow any money from the Bank, preferring to save for things first.

We had good neighbours. Another neighbour was Mr and Mrs Thomas Knowles, at Hollowgate, about a mile away. They had just started farming about the same time as us. We shared many things and helped each other to do heavy jobs. We bought a muck spreader between us. I caught all the moles I had on my own land as well as three of my neighbours and continued mole catching into my retirement.

I had eighty sheep, one milk cow, three calving heifers, four calves, about seventy hens, one hundred chickens, a few ducks and three hen huts. I still worked at the Auction and other jobs for a year. Maggie went to work as well helping in the Auction Cafe, Brian passed his eleven plus and went to Kendal Grammar School, he would walk with Dorothy to catch the bus at the Jungle Cafe. Mary had left school and was working as a dairy maid, at Mr Jennings, of Must Hill, Skelsmergh, so that was a help.

As money came in we bought more hen huts. Hen huts cost thirty pounds and this enabled us to keep more hens and that brought more money in every week. We increased our sheep and cattle slowly. After a few years we started to keep turkeys and dressed (plucked and cleaned) them at Christmas. We did forty or fifty at first, then over one hundred. The most we dressed was two hundred and we had customers for them all. They weighed from eight to thirty-two pounds. We had our own breeding stock and hatched all summer, to get small birds as well as large ones. When it came to the week before Christmas, everybody helped us to pluck the turkeys. The two girls were working at Goodacres Carpets at this time and helped in the evenings or weekends, Brian usually took a week of work, Kit Hawes and Elsie (Maggie's sister) also helped. We plucked them on the first floor of the Salving House and Maggie had the final job of drawing (cleaning the insides out) and tying them. The tricky part, was getting the correct weight turkey to the order required. It was hard work, working from six in the morning

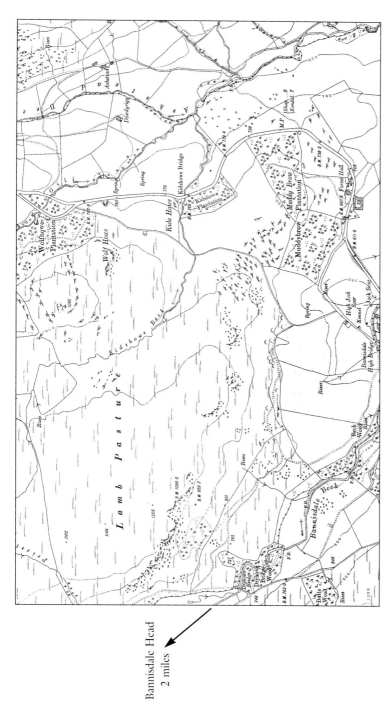

Reproduced from 1920 Ordnance Survey Map.

Bannisdale Head
2 miles

51

Turkeys at Christmas in front of Borrowdale Head.
"A hell of a lot of feathers to take off!"

Me standing with two of my best Masham gimmers at Borrowdale Head.

This was a pen of my best Mashams taken at Borrowdale Head, before going to market.

Taking a break while haytiming at Borrowdale Head in 1958
Me, Maggie, Mary and Brian

until ten or eleven at night. Then usually the last couple of days before Christmas was spent delivering them to our customers.

We also sold hatching eggs, day olds and grown poulets. The money we made from these sales paid for the feed to fatten the Christmas birds. We also had ten ducks and the eggs were used for cooking and my breakfast. The first few years were hard, sometimes I could not afford to smoke. Eggs from the hens paid for the feed for the stock and ourselves for the first few years. We had pigs, we kept two for ourselves for bacon and sold the rest.

Maggie's youngest sister and her husband Reg, came to stay with us in the summer for their fortnight's holiday and they would bring their three young children. The children played in the streams, it was quiet and there was plenty of room for them to run about. Reg helped me with the hay in return for their lodgings and they shared with the cost of the food. Once Reg stayed a week and it rained every day and we got no hay. Brian and Reg went fishing one day after coffee, up Borrow Beck to the source, then over into Crookdale and fished down to Hause Foot. They came home at teatime with ninety-five trout and one sea trout. We had a lot of fish meals! Once when I was fishing, I met two old otters and three young, they soon disappeared, it was no use going any further!

Uncle Tom used to come to help. His wife, Mary, had died suddenly walking for the bus one day, she was only in her fifties. So when he came to live with his daughter, near Kendal, he got a job at Helme Chase as gardener and would come and help us at weekends and holidays, it was nice.

I remember a treat we used to have, from the thick and creamy milk after a cow had calved (about her third milking). Maggie would put this in a dish with some sugar and nutmeg and pop it into the oven until it set. We would have it warm, it was like ice-cream, it was called 'beast pudding'.

I remember the second lambing we had at Borrowdale, it was good weather but we had other troubles. I had six sets of twins in a field, next to the lambers. I looked over the wall and saw five lambs lying dead worried by a fox, enough to waken you up. It had bitten them in the neck and sucked the blood out for its cubs, one lamb off each of the five sets and one of the other set was bitten but it recovered.

The Ullswater Foxhounds came and caught a vixen and eight cubs in Bannisdale, two miles away. A fox does not know it is doing wrong, they are just feeding their young. A fox goes hunting miles away from its cubs, gets food and then by the time it gets back to its hole, the food has mixed with its stomach juices, so that when she regurgitates it the cubs can digest it easier. The cubs have this when they are very young, then they progress to the liver, heart and lungs, and in the end when they are older, they have the flesh and whole lamb. We lost odd lambs some years but not many.

One year in the early seventies, it was a good winter and spring, I lambed one hundred and twenty sheep and never saw a dead lamb, they all lived and were sold in the autumn. The following year I had six lambs stillborn before I had a live one, that is the luck of farming. The field house at Borrowdale was a quarter of a mile away, sometimes in winter I had to dig the doorway out, the cows, six of them, went across the field to drink. The stream was often to dig out, sometimes it was blown full of snow again before the cows got there.

By now we had twenty cows and calves at foot, we also had some we reared on the bucket and one hundred and twenty sheep and a few hundred hens. Some years we even kept a few pigs. I remember when we sold sucklers at first at Kendal Auction, we were getting between £25 to £30 each. These sucklers were calves six to eight months old, reared by their own mothers, normally only the one. By the time I retired I was getting between £150 and £200 each for my best sucklers.

When we first moved to Borrowdale, we did not have electricity, we just had tilly lamps, candles and later on calor gas. We got electricity in 1959, when my landlord paid for the wiring and the cost of putting it in. We couldn't have afforded it otherwise and the interest of the cost of putting it in was added to our rent. The first three years we were in Borrowdale Head, the road was private and me and my neighbour from High House, the farm further up the valley, had to dig it out when the snow blocked it. Once we dug for two days to clear it and a storm came that night and filled it again. Then the Council took it over and they cleared it out with a snow plough. Later we had pylons, water and gas pipes running through.

Both my daughters got married in Selside Church, Mary was first on 21 September 1963 and she married Alan Cowperthwaite from Middleton and Dorothy married Brian Howson, from Barrows Green, on 26 September 1964. One suit did for both, it was a saving. When Mary married, she was the first bride from Borrowdale Head for one hundred years, the last was her great grandmother on her mother's side. The family then moved away ninety years ago and it is remarkable that some offspring should return.

Once when Maggie's great granddad was coming home from Kendal in the 1840's, he had been selling stock and he was riding on horseback through Wolf Howe Wood, where the Jungle Cafe is. The wood was thick and it was dark, someone grabbed his leg, he hit him with his stick, then hit the horse and galloped away. This person knew he would be carrying money. Human nature has not changed much in all the years. The old road over Shap Fell was not far from Borrowdale Head, there are stiles in the walls which join the road wall, they are about three hundred yards from the road. An old man told me, the farmers' wives from Shap took their produce in baskets to Kendal to sell, they followed this track away from the road, out of sight of robbers.

I loved my life in the hills, studying animals, their ways and habits. I once saw a pine-martin in Bannisdale Woods, I think it died in 1947, in the bad weather. I've seen foxes mating in the snow, watched peregrine falcons in the nest, seen the male bird bring a pigeon for the female. She would fly off, he would drop what he had in his claws and she would catch it in mid air, he would go on to the nest while she fed on a rock nearby. I once watched a stag with about twelve hinds, he kept them in a circle walking

A photo taken recently of me standing outside the restored Salving House at Borrowdale Head.
Photo by Phillip Bonney.

My son Brian, taken recently at Borrowdale Head on his quad bike, with Sam on the back.
Photo by Phillip Bonney.

Me and Maggie celebrating my 80th birthday at a hotel near Ulverston.

Me taken recently with Sam, Brian's dog, rounding some sheep up near the yard at Borrowdale Head.

Photo by Phillip Bonney.

57

round them, there was a young stag at each side, when one got too near he would chase it away a few hundred yards. While he was occupied, the other stag ran in and served one of the hinds, it was like playing 'opportunity knocks'!

There were many cuckoos in those dales, there was once five shouting in the trees in Bannisdale yard. We once had a duck sat out on some eggs, we did not move it because it would have stopped sitting, we knew when it was due to hatch. Two days before the eggs were due to hatch the fox went with duck and eggs. It must have passed it many times but it had left them until the eggs were full to get a better feed, it must have heard them squeaking.

When we lived at Bannisdale we passed the long nights pegging mats and listening to the radio. I've made many walking sticks out of hazel, ash and blackthorn, but have never attempted to do tup horn sticks. I have judged sticks many times at Shows but have not shown my own.

When Brian left school at fifteen, he worked on farms in Selside before going as a shepherd to Mr Wilson's at Troutbeck Park, at the top of Troutbeck Valley, not far from Windermere. Ten years later when Mr Wilson retired, Brian came home and did casual work. When I came up to retiring, he did not wish to carry on the farm by himself as he was not married. Levens Hall Estate kept Brian on as shepherd at Borrowdale and it was farmed together with their other farm, Forest Hall, which they had also taken back into their own hands, when Mr Dargue retired.

We retired in 1981 and we stayed on with Brian. It was his home and went with his job. Brian is keeping up the tradition of writing songs and has won more prizes and attended more social evenings than I have. In 1990 Brian received the BEM (British Empire Medal) for his work with sheep, this award going to the ordinary shepherd. Brian continues shepherding the land where I spent thirty-two happy years. He now covers the fell on one of those four wheeled motorbikes. Dorothy has acquired some of my father's skills, she can sketch, paint, wood carve and ice wedding cakes. Mary is also talented and likes doing embroidery.

The house and buildings at Borrowdale are a thousand feet above sea level and it is lovely in spring and summer, being alone with nature. With this extra free time on my hands, I spent some of it catching moles on Shap Fell. I remember the weather was good that year, and in the first week in May I caught seventy-eight in seventeen days, never was I so lucky! Some moles are hard to catch and I tried many ways to outwit them, like pressing the run down with my heel either side of the trap or pushing a ball of grass in. Sometimes this worked, if not I reverted to poisoning a worm but I did not like doing this. You had to obtain a permit from the Ministry of Agriculture before you could obtain the poison from the chemist, and this had to be kept locked safely away. I spent the next seven years at Borrowdale and helped Brian repairing stone walls and other jobs on the farm.

Maggie and I spent our spare time walking up in the Lakes in summer and took many photographs. I also went once or twice a week following the Hunt during the season.

When we later moved to Kendal, I still did a bit of mole catching and I went to catch moles for a friend, but only caught six. I guessed there should have been at least twenty, they must have smelt I had become a 'townie'. Not to be beat, I poisoned them in the end. I still continued with all my other pursuits, took up gardening and helped out friends and relations.

I've worked hard all my life and if I could, I would live it over. I have never travelled far or wanted to. I have in a way, put my thoughts and feelings into my poems and songs and perhaps the last verse I have recently added to my song, 'An Old Man's Tribute to Lakeland', sums up what I feel now.

Now I've grown old and my legs they are tired
I gaze at your beauty I've often admired
When I look to your mountains my heart fills with pain
For I know I shall never climb up there again.

Raymond Wightman   George Walker   Joe Gregg Jim Mason JJ Leek   Billy Brownrigg
(by fence)   (by tree)   (3 men standing together)

Troutbeck Shepherds Meet.   These names are accurate to the best of our knowledge.

Photograph by Joseph Hardman. Supplied by Cumbria Heritage Services, Kendal Library.

CHAPTER SIX

MERRY NEETS, MY SONGS AND POEMS

The original Shepherds Meets and Merry Neets were held in the autumn. Merry Neets started as a 'do' after the annual sheep swapping (Shepherds Meet), usually held at a pub for a meal and a drink, followed by story telling, songs were sung and prizes could be won. In the early years farmers and shepherds attended on foot or horseback but this gradually changed as time moved on. The most noted Meet was at Mardale, at the Dun Bull, before the valley was flooded to become Haweswater, to provide water for the people of Manchester. With the coming of motor transport, sheep swapping has become an every day affair and so Merry Neets have changed into social gatherings at different places.

I only attended one Neet when I was at Bannisdale, as it was difficult with the children being so young. It was different when I was at Borrowdale, the children were older and it was not as far to travel and we would maybe go to about one a month. I have been to Neets at Bampton many times. Every singer had his own songs and when each person got up to sing, you knew what they were going to sing. To start with, I had only one song of my own but I got fed up of singing the same one everytime, so I started to write more songs. I have about twenty songs altogether of my own and have sang all over Westmorland. At Pooley Bridge there is a competition held each year for a new song, with eight or nine competitors taking part each time and I have won this competition several times. At these gatherings there is usually hotpot, or pie and pea suppers, then the wine flows freely and the MC calls to order and the entertainment starts. I still attended these socials after I retired. It was a great time for people getting together for a good night, sing songs and story telling. Denis Westmorland a folk singer, has sung two of my songs and put them on tape, 'Our Johnny' and 'The Langdales'. These Merry Neets still continue yet in the north of the County, though not to the same extent.

Here are some that I have written:-

SOLITUDE

(The first poem I wrote, when I was on my own at Bannisdale and my favourite)

Away from bustle and from crowds
Where the hawk and raven dwell
There stands alone in solitude
Half hidden by the fell
A home of peace where nature reigns
And she alone is queen
A corner of old England
That is so seldom seen

Now city folks can have their films
Bingo and parties gay
But give to me a peaceful vale
When the evening's turning grey
By a good log fire I'll sit and smoke
And rest in comfort there
A home of love and peace and joy
That anyone can share

When the sun sinks slowly in the west
Spreading her golden glow
The deer come down the mountain path
To the pastures far below
A barking fox disturbs the dusk
It echoes in the glen
A scene that never will be known
To many a thousand men

When mountains wear a cap of grey
And the sky's a crimson hue
The wind comes roaring up the dale
Like an express passing through
The noble pines bow down their heads
Above the dancing ling
It carries on through bent and rush
And fairly makes them sing

And when the raging storm is past
And the sun doth shine again
The mountain stands in beauty
Like a bride with a golden train
The silver streams come sparkling down
Bounding o'er rock and stone
How beautiful old nature is
When she is left alone

As here in this dale I sit
My thoughts do sometimes roam
To busy streets and buzzing cars
The place some folk call home
But I'll never be more content
Where ever I may go
Than here in this, our peaceful dale
With nature in the raw

LAKELAND

We talk and we sing about beautiful Lakeland
The hills and the valleys are lovely to see
The Lakes and small tarns shine there like silver
A more lovely picture there never will be.

As you walk o'er the hills in this beautiful Lakeland
The lark and the curlew fly high overhead
On top of a rock end a red stag is watching
And the badger lies low in yon bracken bed.

When you walk on the mountains you look for sly Reynard
A meeting with him is so very rare
But late in the evening you may see him trotting
Down into the valley for a rabbit or hare.

On yon steep rugged hillside the herdwicks are grazing
Their dark freckled coats mingle in with the scree
They're as hard and as tough as the men who have bred them
And herdwicks in Lakeland there ever will be.

We love our dear Lakeland its hills and its valleys
We love every river and every tall tree
The thousands of people who come here to share it
Are spoiling the beauty they've all come to see.

THE LANGDALES

(Where I was born and brought up)

Here's to the Langdales, the place where I was born
Us lads we would gather upon a fine summer's morn
We'd climb right up Wetherlam to see the white hare
Then away on to Dow Crags, Seathwaite Tarn lying there
Back round Grey Friar, across Cockley Beck Fell
Then back home down Wrynose, the road we knew so well.

We'd start off up Lingmoor, climb out by Busk Pike
Then right through by Blea Tarn, up Blake Rigg we'd hike
Around Pike O'Blisco and Crinkle Crags too
Across to Bowfell with the pikes in full view
We travelled back down t'Band and in by Stool End
Down the valley bottom our way we did wend.

We'd set off from Colwith, climb onto Brow Fell
Away on by Black Crag to Iron Keld
By the beautiful Tarn Hows, down St Mary's Glen
Across lovely Yewdale and back home again.

Now it's back to the Langdales to finish my song
They are nearly all strangers, many families have gone
But they can't change the beauty it will always remain
And I have but one wish, to be born here again.

MOUNTAIN GLORY

These are our mountains where we love to roam
This corner of England that we call our own
Those lands may have splendour way 'cross the sea
But beautiful Lakeland is homeland to me.

Chorus
These are our mountains and this is our glen
We've hunted and fished here both as boys and men
This land of our fathers we love to the end
Where each man's a sportsman, a brother, a friend.

We've scrambled the mountains from valleys below
Led on by the huntsman and loud tally ho
On reaching the high tarns where rocks all abound
The echo returns there many times all around.

Chorus

The great golden eagle is hovering on high
His golden plumes flashing against a blue sky
This majestic creature so great and so rare
Is searching for foodstuff - a rabbit or hare.

Chorus

And as we descend though reluctant to leave
The height of the hills is so great to achieve
The valley's below us, above is the lark
Through bracken and crags - getting home before dark.

Chorus

The folks from our valleys are honest and true
To help you there's nothing that they will not do
When someone's in trouble they all rally round
They won't kick a man when he's on the ground.

Chorus

And if they stop hunting at some future date
We've songs and we've stories that we can relate
We'll sit by the fireside when evening turns grey
And bring back those memories they can't take away.

Chorus

AN OLD MAN'S TRIBUTE TO LAKELAND

Chorus
One hill at a time dear Lakeland
That's all I am able to do
The pleasure I've had since I were a lad
Wandering o'er you
Hunting and climbing on your mountains sublime
If Heaven's like Lakeland I'll go anytime.

Of't I remember when I was a boy
I wandered your hills and it filled me with joy
Looking for birds' nests and fish in your streams
Oh beautiful Lakeland the land of my dreams.

Chorus

And in the winter when the skies are all dull
There's snow on your hills or your rivers are full
But when the spring comes the sun it will shine
Over all Lakeland, this haven of mine.

Chorus

I've hunted your mountains with all the fell packs
I know every borran and all Reynard's tracks
I remember my hunting with story and song
I hope you're still hunted long after I've gone.

Chorus

Now I've grown old and my legs they are tired
I gaze at your beauty I've often admired
When I look to your mountains my heart fills with pain
For I know I shall never climb up there again.

Chorus

THE LAKELAND RAMBLER

I've walked over Scafell and back around Bow Fell
Over Gable and Great End as well
I've done Langdale Pikes and up Pavey Ark grikes
And finished off down Langdale Fell
I have my big walking boots on and my stockings come up on my knees
If my shorts were a little bit shorter
My rucksack would swing in the breeze.

Chorus
I'm a rambler, I'm a rambler round Lakeland's fair way
I often do twenty miles a day
I go like a wild deer on Monday
But I'm knackered by Sunday.

There's girls from the city and they are so pretty
They come to the mountains to roam
They couldn't dress better, they're in shorts and a sweater
And they never want to go home
They come up the path and pass by me
That's something that I do not mind
I'm willing to follow on after
I get a much better view from behind.

Chorus

When you go for a walk you meet all kinds of folk
Some from the countries abroad
There's old and there's young and some bad in the lung
They'd be far better off on the road
I met up one day with a Yankee, he spun me a hell of a yarn
How he's stood on the top of Helvellyn
And piddled right into the tarn.

Chorus

When I go to the hills, I should take some pep pills
To help me to get to the top
When I get up there in the clear mountain air
I'm just about ready to drop
Some days I like to go fishing, it's peaceful and makes a nice break
As I sit on the shore of Ullswater
And dangle my worm in the Lake.

THE LITTLE LANGDALE LAD

(Dennis Barrow was born and bred in Little Langdale and our grandparents were related)

There is a Little Langdale lad, Dennis Barrow is his name
He's Huntsman to the Ullswater: that's where he won his fame
He catches foxes on our fells, this makes the hunters glad
And that is why they follow him, this Little Langdale lad.

He started with the Coniston Pack, Anthony Chapman showed him how
He taught him to feed the hounds and how to skin a cow
He found out when the scent was good also when it was bad
And now those foxes fear him, that Little Langdale lad.

Fra Kirkstone Pass to Pooley Brig', on either side of the lake
He keeps those foxes on their toes, they have to stay awake
Up Mardale Head, round Sleddale Fells and back again to Shap
His voice it makes the mountains ring, that Little Langdale chap.

He went to Meet at Dockray pub and had a gay good day
A fox it went among his hens while he was away
He came back and saw them lying dead, he said, "Now that looks bad"
And when everybody laughed at him it made him very mad.

He holed a fox on Sleddale Fell, his hunting pals were there
There was one with a lot of whiskers on and some with gay long hair
They were listening down the hole for t'fox when it bolted with a rush
He collared Ronnie Bell by t'beard and thought he had the brush.

Now hunting stops in early May when weather gets too warm
And Dennis goes a shepherding upon a neighbouring farm
When the weekend comes around he does not go so far
He puts his little apron on to wash up in t'snack bar.

We sent him home from Langdale, he had a touch of flu
Denise said, "You must go to bed, I'll go and have a doo"
'She off with stick and horn and whip, and went up Reg's wood
And caught a fox in High House Wood, now didn't she do well.

My hunting days are getting less as time goes quickly by
But I can still get up a hill .... if it's not too high
When I think about the days we've had it fills my heart with joy
How glad I am I've hunted with that Little Langdale boy.

Dennis Barrow of the Ullswater Foxhounds.

THE A6 ROAD

(Written a few years ago when I was asked to sing at a Social Evening at Selside)

One day I thought I'd write a song, about a place I know
So I got my pen and paper out, and then I had a go
I did not have to use a map, it's a place that I know well
I wrote about the A6 road, from Kendal to Shap Fell.

Chorus
So we'll travel along the old A6, from Kendal to Shap Fell
It is a long and winding road, it's one that I know well
It used to be a busy road but it's not like that today
The traffic still goes north and south, upon the motorway.

Jack Tallon lives along this road, at a place they call Watchgate
I've known him over sixty years, he is a right good mate
I know he likes to have a pint, and tell a few good tales
But Violet always goes with him, to keep him on the rails.

Chorus

The old Plough Inn's still standing there, it's served the public well
There's many a traveller had a pint, to help him climb Shap Fell
They've modernised it quite a lot, they've done a right good job
When I was young I could go in there and get drunk for a bob.

Chorus

Just up the road beyond the Plough, our eyes turn to the right
We look across at Cooper House, where the bomb fell on that night
One whole family wiped out, it was a bitter blow
Why it had to drop right there, no one will ever know.

Chorus

Up to the left is Bannisdale, I lived there eleven years
The family were young when we were there, we'd laughter and we'd tears
Living up there in solitude, away from all the strife
I often look back on those years, as the best in all my life.

Chorus

There is a caravan park now, up there in Wolf Howe Wood
It is on the corner, where the Jungle Cafe stood
This cafe was a noted place, lorries called from near and far
They always got a right good meal, but it was not five star.

Chorus

There's a person up at Hollowgate, who speaks on the radio
She tells us every morning, whether it's going to rain or snow
One day she said, "It's a roaring, it's blown away all the mist
My washing is out on the line, and knickers are in a twist."

Chorus

The Leyland clock stood on yon hill, in hail or rain and snow
Each driver he would check his watch, to see if it was slow
It always seemed to go quite well, but I never heard it chime
Lenore she used to wind it up, for about three pence a time.

Chorus

When that whisky wagon ran off the road, at the bottom of Huck's Brow
There was glass and bottles everywhere, there was an awful scrow
There was whisky flowing down the beck, this made the salmon rise
I saw one go past Borrowdale Head and it had bloodshot eyes.

Chorus

When the Scottish weekend came around, down to
Blackpool they would go
They all went to see the lights, then on to see a show
On Monday they would all return, but the journey it was slow
I seen thirty buses climbing Shap, all in one long row.

Chorus

So now I'll finish with my song, from the summit of Shap Fell
I've told you many stories, there are more that I could tell
I do not know which way is worst, coming up or going down
There's many a driver sighed with relief, when he's reached Kendal town.

A line of buses making up Shap Fell in the 1950's. Glasgow Fair weekend, and the Glaswegians are making their way home after being to Blackpool.

BORROWDALE HUNT

Ya morn when it was barely leet. Ha way! Ha way!
There was sna on ivery hill I seet. Ha way ma lads. Ha way!
Dennis set off from Borrowdale Head
Where he was staying with Maggie and Fred
He'd been loth to leave that feather bed. Ha way ma lads. Ha way!

We gat him up in gay good time. Ha way! Ha way!
Afore eight o'clock did chime. Ha way ma lads. Ha way!
He cleaned his hands and laced his shoes
And listened to the morning news
Which way to ga we had to choose. Ha way ma lads. Ha way!

On t' High House Bank we let them ga. Ha way! Ha way!
We followed on amang the sna. Ha way ma lads. Ha way!
In t'building nick they struck a drag
Thad led us on to Jopson Crag
Those gallant hounds their tails did sway. Ha way ma lads. Ha way!

A fox got up and reet down t'Breast. Ha way! Ha way!
Those hounds had spoiled his morning rest. Ha way ma lads. Ha way!
On t'bank Breast they ran him well
Across t'back fields and up Shap Fell
And ower to'Knott they went like hell. Ha way ma lads. Ha way!

Across t'Breast High without a check. Ha way Ha way!
They turned him in for Borrowdale Beck. Ha way ma lads. Ha way!
Then up the valley on t'cart track
The hunters said "He's coming back"
They saw him climbing up yon slack. Ha way ma lads. Ha way!

Back on t'Bank he thought he'd ga. Ha way! Ha way!
His brush was trailing in the sna. Ha way ma lads. Ha way!
But on t'Bank end they turned him round
Cross Crookdale Beck he vaneer drowned
And under t'Shap road he went to ground. Ha way ma lads. Ha way!

Reg Dixon loosed his terrier 'Rock'. Ha way! Ha way!
It gave that wily fox a shock. Ha way ma lads. Ha way!
It bolted out reet down t'beck
And Barmaid* collared it be t'neck
Folk out on t'road said "Eee by 'eck". Ha way ma lads. Ha way!

(* Barmaid - name of a hound)

OUR JOHNNY

(In memory of John Richardson, Huntsman of the Blencathra Hunt)

Now gather round you hunters, a song I'll sing to you
About a Lakeland Huntsman who was so good and true
His name was Johnny Richardson, he always did his best
But Johnny he will hunt no more, for he has gone to rest.

Chorus
From Saddleback to Skiddaw, up Newlands bonny dale
In Borrowdale and Wythburn Head, down St John's-in-the-Vale
That's where we went a hunting with Johnny and his hounds
On every day we spent with him our pleasure had no bounds

Up Mungrisedale's bonny valley, on Carrocks's rocky fell
Up Swineside Head and Skiddaw House, those hills he loved so well
And down in by the Dash Falls and on yon Skiddaw side
Those foxes had to go like hell and find somewhere to hide.

Chorus

Now Johnny was a soldier, he was captured by the Hun
But they could not hold him, he was often on the run
He was as sly as 'Cathra foxes he hunted on the fell
We've listened to the stories he sometimes used to tell.

Chorus

Our Johnny had a gallant pack, some forty hounds or more
In the local show rings he won rosettes by the score
He was a noted breeder his advice was often sought
He held the season record with the foxes he had caught.

Chorus

One morning up at Caldbeck, the anti-hunters were around
But Johnny turned out with his hounds and firmly stood his ground
He said "Hunting's always been my sport and I will carry on"
And Barry guides those gallant hounds now that he has gone.

Chorus

When Johnny was a bouncing bairn he won a baby show
The spirit of competition lived on with him you know
For at our agricultural shows, keen interest he did keep
He was often called upon to judge the Herdwick sheep.

Chorus

A woman was out hunting, she said his hounds they were no good
They couldn't even smell the fox, but she said that she could
Johnny looked at her and said "With you I don't agree
Next time I want a litter of pups I think I'll breed off thee!".

Chorus

On Skiddaw's lofty mountain on a wet and windy day
We still hear the sound of Johnny's voice as he shouts "Whoo get away"
We know that he has gone away and he will no come back
But we have those lovely memories of Johnny and his pack.

Johnny Richardson on the right with Phil Davidson, Joint Master of the Blencathra
Foxhounds, at the Salutation Inn, Threlkeld.

A TRIBUTE TO JOHN

(On the retirement of John Nicholson, Huntsman to the Lunesdale Hounds)

We'll just drink a toast to a Huntsman so grand
His name will go down with the best in the land
Like John Peel and Joe Bowman, such great men as these
John is a man always ready to please.

Chorus
Tally Ho! Tally Ho! Tally Ho!
Hark forrard good hounds. Tally Ho!

John was bred off stout farming stock
Away up on Kirkstone he tended his flock
When he gathered them up on the red screes so high
A smile crossed his face when Reynard passed by.

Chorus

Now John he grew up, he lost interest in t'land
He went hunting with Chappie and he thought it was grand
He said, "Now this hunting's as good as it sounds"
So he joined Walter Parkin and his fine Lunesdale Hounds.

Chorus

John he was keen, he soon learned the rounds
In a few years he could manage the hounds
So when Walter retired he took over the pack
He went on from there and has never looked back.

Chorus

Now John's bred some good hounds for finding the scent
And won many prizes at shows where he went
When they're out in full cry it's a wonderful sound
They've killed more in t'open than they've done underground.

Chorus

This hunting has altered since John took command
They've bleepers on terriers and CB's in their hands
They have all these fine toys in this modern trend
But poor Reynard's nowt modern his life to defend.

Chorus

The job as a Huntsman it isn't all glee
For all these Committees they cannot agree
They say, "Do this here and do that there, John"
But he just said, "Hi-Hi" and then carried on.

Chorus

Now John is retiring, we wish him the best
He can put up his feet and have a good rest
But we'll always remember those days on the fell
With John and his foxhounds that he loved so well.

Chorus

John Nicholson talking with the Hound Judge at Eskdale Show, in the early 1980's.

ANTHONY BARKER'S HARE HOUNDS

(Anthony Barker was Huntsman and Master to the Hare Hounds. They were called,
'The Windermere Harriers', and they were kennelled at Patterdale)

Anthony Barker went to Forest Hall
It was the opening Meet
He said, "Now lads if we shape oursel's
We might catch yan be neet".

Chorus
A hunting we will go my lads
A hunting we will go
For Barker's there to catch a hare
He'll always have a go

We hed our coffee and set off
We clum right up t'wood
But when we got to Muddy Brow
The scenting wasn't good.

Chorus
They put yan off on Muddy Hill
It went around Wuf Howe
Those hounds were about as wild as hares
They made a bloody scrow.

Chorus

They steadied up across the beck
And up among yon bent*
Across t'back half and down yon rocks
For t'Horseshoe Wood they went.

Chorus

Then on t'Lamb Pasture bottom
Over t'Intakes they did go
Across t'Pond Field and through yon wood
Aback a Forest Ho.

Chorus

We lost the scent on t'main road side
So we tried them back again
Anthony said, "We'll not stop long
I think it's going to rain."

Chorus

He tried them back on t'pasture breast
And turned in down Ash Howe
They went like hell down t'Intake trod
He was making straight for t'Plough

Chorus

We gat to t'Plough and hed hotpot
And washed it down wi' beer
Anthony looked at me and said
"We're much better off in here!"

Chorus

(* bent - rough grass)

A caricature done of Anthony by Mr Billy 'Wilk' Wilkinson of Keswick, which is framed and
hangs on the wall in Anthony's home. By kind permission of Mr & Mrs Barker.

A TRIBUTE TO A HUNTSMAN

(In *memory of Anthony Chapman of the* Coniston Hounds)

I sing you a song of a Huntsman
A man we all thought so grand
When he hunted the Coniston Foxhounds
He was the best in the land
Round Scandale, Rydal and Grasmere
He hunted with many a friend
And on every fell that he hunted
He knew every rock and hill end.

Chorus
So we'll all bid farewell to our Chappie
No more a hunting he'll go
But as long as we live we'll remember
The sound of his view Tally Ho!

Those Coniston Hounds they adored him
In his terriers he took great pride
When he set off to hunt in the morning
Everyone would be at his side
Through all the years we have hunted
Our pleasure it knew no bounds
When we hunted with Anthony Chapman
And his gallant Coniston Hounds.

Chorus

He was one of Lakeland's great Sportsmen
Like those whose songs we have sung
But the songs that we sing about Chappie
Will be in our own Lakeland tongue
He was respected by each one who knew him
And where he'll be now is quite plain
I know I will give a great "Hallo!"
If somewhere I meet him again.

Chorus

Anthony Chapman front right, walking up the hill towards the Queens Head, at Troutbeck.
Whipper-In, Chris Ogilvie in rear of picture.

THAT CUNNING OLD FOX

On a hill over there stands a cunning old fox
His tail hanging low in the breeze
We've hunted that fox on many a day
With his cunning, those hounds he would tease

Chorus
And we've hunted that cunning old fox
Till those hounds they were weary and lame
We will look for that cunning old fox
And we'll hunt him again and again

Chorus

That cunning old fox up there on yon hill
He's the master of every good hound
He will run any day from daylight till dark
And he knows every inch of the ground.

That cunning old fox rears a litter each year
Somewhere up there on the scree
They are sturdy and strong, just like their old dad
And almost as cunning as he.

Now that cunning old fox he likes hens and ducks
And sometimes a lamb he will kill
He'll go down to the fields at the break of the day
And take one for his cubs on the hill.

That cunning old fox the day it did come
When his brush he no longer could save
He succumbed to the hounds on that rugged fell side
And we thought of the sport that he gave.

And we hunted that cunning old fox
Till those hounds they were weary and lame
No more will we look for that fox
For we know we won't find him again.

Now that cunning old fox no more will he roam
His remains lie up there on the rocks
There'll be many a day when we pass this way
We will think of that cunning old fox.

Not again will we hunt that old fox
Till those hounds they are weary and sore
It's farewell to that cunning old fox
For we know we will hunt him no more.

BORROWDALE HUNT 1981

Now all ye hunters I'll tell you a tale
Of how we went hunting in wild Borrowdale
We set off up t'Bank, the morning was good
But ne'er got a drag until we reached Robin Hood.

Chorus
Tally Ho! Tally Ho! Tally Ho!
Hark forrard good hounds. Tally Ho!

We dragged way up Seal Green and round White Howe End
And down Mowdy Rake the Hunt it did wend
Bold Reynard got up, he nivver looked back
Right through be t'tarn and West Nab was his track.

Chorus

On t'Forest Breast as hard as he could
He crossed through be t'sheepfolds and clum Robin Hood
Cross Crookdale Beck, round Li'le Yarlside he went
Right for Wet Sleddale, he seemed to be hell bent.

Chorus

They turned him down Wasdale they were running him well
He crossed by t'young planting and clum for Shap Fell
Right ower t'Bank across Borrowdale Beck
Then right out by West Nab with never a check.

Chorus

Away ower to White Howe, round Bannisdale Fell Head
He ran right down t'skyline, along t'water shed
Down Dry Howe Wood he vaneer flew
And up on to Doddin to a hole that he knew.

Chorus

The hunters got there put a terrier in
And afore very long there was a hell of a din
Reynard dashed out they all gave a shout
They were all gay capped when another came out.

Chorus

All t'hounds went with t'first yan, they rushed him in t'bank
And killed him in t'bottom where t'rushes are rank
Young Robert Hayton was stood up in t'rocks
He said this is better than showing game cocks.

Chorus

The other fox had a good lead by now
He clum up Doddin End and ower t'White Howe
Round Bannisdale Head down the wood he did go
Across Bannisdale Road and up Capple Flow.

Chorus

Keeping up with this Hunt was getting hard wark
It had come in gay misty and was fast getting dark
Ower t'forest top and down Willy Gill
Those hounds were running and meaning to kill.

Chorus

Across Shap Road and up Ashstead Fell Breast
Rob and Harold were arguing which hounds were ganin first
Right across Borrowdale in a straight line
And we gat them fra Roundthwaite at round half past nine.

Chorus

GREENHOLME HUNT

Now all of you hunters just gather near
I'll sing you a song that you don't often hear
You hear quite a lot as you go on your way
But you seldom hear one when the fox wins the day.

Tally Ho! Tally Ho! Tally Ho!
Hark forrard good hounds. Tally Ho!

One Saturday morning to Scout Green we went
With John Nick in charge and to kill one we meant
We tried away forrad towards t'shooting box
And Bellman* gave mouth he'd got wind of a fox.

Chorus

This drag led us on to Bretherdale Bank
And there avout t'heather grows rank
Bold Reynard got up, and away he did go
The sound in his ears was a glad Tally Ho!

Chorus

Away round t'Bank End he made for Breast High
Then turned right down t'pasture with t'hounds in full cry
Than up t'valley bottom he made for Shap Fell
"I'll give them some climmen" he said to his sel.

Chorus

When he got on to t'top he off like a shot
Away over Greenside and down in by t'Knott
Ower Borrowdale Beck and up Ashstead Fell End
Up among yon rocks his way he did wend.

Chorus

He couldn't reach t'skyline although hard he tried
So he turned off through t'Breast - kept in Whinfell side
And on by t'Slape Stones and into Grizedale Lotment
Across t'bottom o'Mabbin right down hill he went.

Chorus

He thout of yon earth holes in Harrad Lot
He was in suck a hurry he couldn't find t'spot
So he went on down t'Sled road with t'hounds pressing hard
And gat ower t'wo into Jack Handley's yard.

Chorus

When t'hounds came in seet he up and away
Crept into t'barn thout he'd hide amang t'hay
He set up his back and stood on his guard
He crept through a hole and jamp down into t'yard.

Chorus

This gave the hounds a hell of a check
Sez he to hissell, "I've saved my old neck"
Than up on t'Poor Lot* in strang spot* he lay
And had to be left there for some other day.

Chorus

We've many a good hunt and the fox wins the day
It wouldn't be sport if it wasn't that way
There's lots of non-hunters may laugh and may scoff
But a stout fox like this yan is fit to breed off

Chorus

When hunting the fox just give him a chance
For the poor la'al divils they nobet live yance
We don't give a damn what he's done in his day
Our Dalesmen and Huntsmen will give them fair play.

Chorus

(* Bellman - name of a hound, *strang spot - strong hold
*Poor Lot - piece of land where the rent paid goes to charity)

LONGSLEDDALE HUNT
(Fictional)

It's a fine hunting day, as bonny as May
The hounds have come to Longsleddale to stay
The guns have been out but they have got nowt
But I think we will get one today.

Chorus
So we'll all go a hunting today
Up Longsleddale we'll wend our way
On Swinklebank Crag we'll be sure of a drag
So we'll all go up Sleddale today.

On Docker Nook rocks we tried for a fox
There'd been one about we could tell
But the scent was not strong for he'd been gone long
And we finished on Nether House Fell

Chorus

Away on we tried just in Kentmere side
Across Stile End Road we all went
And on Shipman Knotts, one of best spots
A fox got up amongst bent.

Chorus

Across Powte Howe grassing and through by Ull Stone
And right on to top of Nan Bield
As he clum Harter Fell he said to his sel'
My brush to those hounds I wont yield.

Chorus

Twice around top with never a stop
Then set off right down Sleddale Fell
He went in below settle earth the place of his birth
And it was his death bed as well.

Chorus

It was not very late so we crossed by fell gate
And right along Buckbarrow top
When we got to Hartmane they put off again
And there was not a hound that did stop.

Chorus

Over Grey Crag top to Borrowdale Moss
Then onto Bannisdale Fell
On that plain ground there was every hound
Running and screaming like hell.

Chorus

Away on down to Dry Howe Fell
Where scenting was not quite as good
But they went over top at a bit slower pace
And then dropped down into Yewbarrow Wood

Chorus

He went ont' fell bottom just above't Church
And at Middale Drain went to ground
This was the end of a good hunting day
And do you know we had every hound.

Chorus

Now this song is all fiction, I think you all know
But I thought it a bit of a joke
You can mix up your words to make them all rhyme
But you can't con those true hunting folk.

Chorus

If this hunting should end, we will have to pretend
And write some more songs just like this
For those nights out and sing-songs we had in our time
Is something that we're going to miss.

THE HUNTERS

We like to see the picture of those hounds upon the fell
The hunting horn that wakes the dawn, the sound we know so well
And then we see a shadow move up there among the rocks
Away he goes in his brown coat, the stout and wily fox.

Chorus
But most of all we long to hear those hounds going in full cry
For in their tones they seem to say, "Bold Reynard you must die"
He scrambles up among yon rocks and across yon screes so grey
And goes like hell to save his sel' and live another day.

It's in the early morning when the dew is on the ground
We stand and watch in wonder at the wisdom of the hound
They'll find the scent of Reynard and drag him to his lair
The music of those gallant hounds is far beyond compare.

Chorus

Now when the hunting's over to a sing-song we will steer
We'll sing of hounds and huntsmen who are no longer here
We'll sing about our mountains and our valleys green below
And mingle in the friendship that only dalesmen know.

Chorus

OLD DARKY

The Meet was at Harrad* yan wet misty morn
Whinfell beacon re-echoed the sound of the horn
We cast off on t'Lotment and cast a cold drag
And Reynard broke cover up on Mabbin Crag.

Chorus
Tally Ho! Tally Ho! Tally Ho!
Hark forrard good hounds. Tally Ho!

He up and away he vaneer flew
For Bellman was there and running to view
Along came the others without a false call
Right down cross Shap Road and through by Forest Hall.

Chorus

Right ower t'Lamb Pasture he made it his track
Then through by t'Dry Howe, he daren't look back
Away out by t'Black Crag to Bannisdale Wood
And down into Sleddale as hard as he could.

Chorus

By now all the hunters were left far behind
For both fox and hounds had gone like the wind
Away up into t'mist Johnnie shouted the way
"They're making for Dry Howe," he heard a voice say.

Chorus

Some set off down t'fell they vaneer ran
To get into Bannisdale that was their plan
"They'll ga right to Sleddale," said Walter , "I knaw"
"Come on then," said Arthur, "We'll ga round in t'car."

Chorus

Mrs Dixon at Yewbarrow was just brewing t'tea
She looked out of t'window and Reynard did see
She down with that teapot and out she did run
Says she to hersel', "He looks about done."

Chorus

He went out cross t'meadow and turned back up t'Breast
With hounds gaining hard he had no time to rest
He turned again down t'wood towards Nether Bower
And Trimmer pushed forrard and he rowled him over.

Chorus

This was the end of a splendid day
His head hangs on t'wo at Bank House they say
There was a bit of discussion but I dar mak a bet
That it was old 'Darky' Pont* said, "We'd never get."

Chorus

(*Harrad - Harewood *Pont - Ronald Hayton)

MEMORIES OF WILLIE PORTER

(Willie Porter was Master and Huntsman of Foxhounds at Eskdale and Ennerdale when I was a boy)

I'll sing of the Eskdale and Ennerdale, one of the best in the land
I've hunted with three generations, Willie, Jack and Edmund
I have lots of beautiful memories, some of them I'll sing to you
They are all about Willie Porter, and every one of them true.

Chorus
We hunted those far western mountains, from Great Gable down to the sea
The great rounded dome of the mighty Black Combe and the rugged Wast
Water scree

I remember when I was a schoolboy hunting upon Lingmoor Fell
I harked on some hounds that were running heel way and Willie came up and played hell
He said, "Come here young fellow, and listen to what I do say
A fox niver goes down hill in a morning, it always goes t'other way."

Chorus

Once Willie was walking to Langdale down the road just below t'Bield
When the Coniston Hounds went by with a Hunt, over the wall in't next field
Willie turned to his foxhounds and uttered only one word
The Coniston went on with their rattling Hunt, and not one of those forty hounds stirred.

Chorus

Those quarry men loved Willie Porter, a good hunt they all liked to see
They would follow the horn on a Saturday morn, and their clog calkers rattled on't scree
And when the hunting was over, back to the pub there to sing
They all had melodious voices, they made those old pub rafters ring.

Chorus

One Good Friday I went hunting, on't top of old Wetherlam
I'd a young hound on't lead called, 'Crafty', it was pulling so I let it gang
It soon put off a white hare and away they both went like hell
Across Swirrel House up among yon rocks and onto top of Brim Fell.

Chorus

We watched that hound hunt that white hare, she'd make a good hound it was plain
Willie came up and said who let that thing off, and I got played hell with again
Us lads we adored Willie Porter with him on the fells we would roam
He was so kind if we got left behind he'd show us the nearest way home.

91

Chorus

One wet misty morning in Eskdale, there were hunters up from the south
One looked a canny little fellow until he opened his mouth
He said, "Where are you loosing this morning, I don't think it will be much cop."
Willie said, "I'll put some hounds in't bottom of wood and t'other half in at top.

Chorus

This little fellow said, "That's a tall story, who are you trying to kid."
Willie stood up on this rocky hill end and what he said he'd do he just did
This fellow stood there in amazement, he'd never seen that happen before
He kept his gob shut for the rest of the day and we never saw him no more.

Chorus

I have many memories of Willie and his Eskdale and Ennerdale Pack
Edmund his grandson now hunts them, in the footsteps of his late father, Jack
Many school mates with whom I went hunting, have quietly gone on their way
They've all gone to meet Willie Porter and I know I shall join them some day.

Chorus

Willie Porter on his 70th birthday, standing with Arthur Irving and hounds of the Eskdale and Ennerdale Foxhounds.

EPILOGUE

FAREWELL

Fred and Maggie are sadly no longer with us and I would like to share with you the last year, as I know it, of Fred's life. His book gave him great pleasure from the very start, the compiling, finding suitable photographs and he was continually adding new stories right up and during the printing stage. Indeed, even then, he came with a large scrap book which he had found full of wonderful cuttings from old newspapers. Miller Turner (our printers) did us proud and when the first books were finished, Fred and I went to collect them on Friday 21 November, 1997, around 5.30 pm. We were over the moon. Mike Miller and the young girl, Vickie Hunter, who had done the typesetting and all the work for us had stayed late so they could meet Fred. We were all delighted, books were signed as gifts and photographs taken of the memorable occasion and Fred was shown around.

We had arranged publicity: writing to newspapers, magazines, local television and radio. Everyone responded and gave us glowing reviews. They all loved Fred for however short a period they had with him. Anne Hopper from Radio Cumbria came and interviewed Fred at his home in November and liked him so much she came back in February. For one week during her programme she played excerpts of Fred's interview. We had the pleasure of meeting Eric Wallace from Border Television and his cameraman, Paul Allonby from Ambleside. They spent most of a day with Fred, both at his home in Kendal and shoots on the farm. It was rushed back to Carlisle to be shown that evening on Border News and Look Around.

We had a thousand books printed to start with and a book launch at Kendal Library, on Saturday 6 December, 1997. Many thanks to Sue Rochell, head librarian and her staff. We wanted it there so that it was easily accessible to as many people as possible. The librarians, Jackie Fay and Sylvia Kelly, did us proud with a fine display and Fred's beloved country folk voted with their feet. We sold a hundred and forty-four books in two hours, Fred was quite overwhelmed with it all and wished he had signed more beforehand. His one regret was that he did not have more time to speak to the many friends and relations who turned up, queuing to see him. It was an extraordinary morning. The books we had left unsold were taken by the Library and were put into Libraries at Kirkby Lonsdale, Sedbergh, Windermere, Ambleside and into the mobile library for sale up and down the valleys. Book shops who had not got copies of his books were phoning asking to stock it and local wholesalers bought for wider distribution. We had a daily post load requesting books and our Post Office was kept busy weighing and sending books across the country.

Whenever I called round to see Fred; he was continually packed out with visitors, he always did have a lot of people calling to see him but they had now quadrupled. He sold many books himself and I used to joke with him, not even the taxi drivers were safe! Everybody he seemed to come in contact with went away with a book. Sadly, during all this Maggie was in Oakdene Nursing Home, Kendal Green, Kendal and was too ill to partake of all that was happening. Fred went up most afternoons and if he didn't, Mary or Dorothy or another relation called to see Maggie. He had read the book to her and said she had enjoyed it. On Sundays the family got together for dinner and Brian would come and Maggie would be taken out for the day.

The book sales were going so well that a rushed order for another thousand books was put in on the Monday after the book signing, as most of the first books had now gone. We had never seen a run up to Christmas like it, the books were still selling thick and fast. What a Christmas! His books went all over the country, as well as abroad to America and Australia. In January, Fred was poorly, he was a diabetic and his sugar level was too high, as well as having fluid in the lungs, so he had to spend a week in the Westmorland General at Kendal. He came home and was carefully looked after by nursing staff and family alike. He loved his greenhouse and garden and when spring came he could often be found pottering away outside. His legs were bad and he was frail, but he never lost his sense of humour and love of company and that lovely 'twinkle in his eyes'!

In February, I entered Fred's book in the 'Lakeland Book of the Year Awards 1998', which the Cumbria Tourist Board held, in conjunction with Hunter Davies. I had to send in four books in February and these were read and judged, with four main prizes for different categories and two runners up for each. The judges were Hunter Davies, along with Eric Wallace of Border Television and Bob Smithies, former Granada Television Presenter. They had fifty-three entries, the most books ever entered since setting up in 1984. We got an invitation to attend the awards presentation and charity luncheon on Tuesday 9 June, 1998 at the Derwentwater Hotel, Portinscale, Keswick. Four of us went along, Fred, Dorothy his daughter, John Marsh and myself. It was a lovely day and a joy to meet so many writers. Fred came runner up in the Barclays Bank PLC section, for the best book on people. We were all thrilled and he was photographed along with all the other prize winners and received a signed certificate, which he hung up proudly on the living room wall of his home together with all the photos of his family.

Sadly, Maggie passed away peacefully on Monday 20 July, 1998 and Fred was heartbroken. There was a large attendance at the funeral held at St John the Baptist Church, Skelsmergh. There was no parking at the church or in the narrow lane running up from the main A6. The main road verges had cars parked on both sides and I had to walk from the Burneside turn-off nearly half a mile away. The church was full to overflowing and I stood at the back, though others were not so fortunate had to stand outside. The vicar quoted passages from Fred's book. David, a grandson (Mary's son) read, "Death is but a pause," by Marian Hayton (a friend of Maggie's from the Dialect Society) and Kevin another grandson (Dorothy's son) read a passage from the bible. I think the heart went out of Fred that day and he was bent double over the grave, I couldn't bear to watch so turned and walked away.

Fred tried to come to terms with Maggie's parting and tried to continue on as best he could. His family and friends were a comfort but could not make up for Maggie whom he had cherished dearly. He continued growing his plants and sharing them by giving away to friends. Alan his son-in-law had taken him to see a hunt and he had sat and enjoyed it from the car as he was very lame and occasionally had dizzy spells. About this time, an independent film company from Bristol, contacted him twice about possibly interviewing him for a four part documentary on country life to be shown on television on Channel 4 but it never materialised. I don't think Fred was well enough. Fred thought of doing another book as he had found the large amount of photographs in the summer for which he had previously looked. There really were some lovely photos and the second book was to have been called, "Further tales from a Westmorland Shepherd'. It was not to be, he never put pen to paper again, bar writing one poem which he wrote and dedicated to Maggie, after her passing. Here it is:-

ODE TO AN ANGEL

Goodbye, God bless and keep you Maggie
A perfect wife and mother
You cared for me for sixty years
There never was another
I know some day we'll meet again
In heaven's pleasant land
Then we will be together again
Walking hand in hand

When I look back on our married life
And all those happy years
All the love we shared together
I can't hold back the tears
The family we raised together
I'm proud to call our own
They are a blessing sent for me
Now that I'm alone

As every lonely day goes by
You're always in my mind
I do feel so lonely now I'm left behind
I always have been true to you
And so I will remain
Looking forward only to the day
When I see you again

Fred

I printed many copies of this poem for Fred, I think he handed over two dozen out to friends. I recall him telling me that he had made his doctor cry after he had shown it to him at the surgery!

In the last week of his life he did not move from his house and had his legs up resting on a footstool. His family came and stayed with him during these last few days in November 1998 and on Thursday 12 November, 1998 when his son, Brian was with him, he turned unwell and an ambulance was called to take him to Westmorland General Hospital in Kendal, which was just a mile away and where he had been many times before for check-ups. When they were carefully putting Fred into bed he was joking with the nurses, his family was with him and when he put his head on the pillow he just closed his eyes and slipped away.

Fred was buried on the following Thursday 19 November, 1998, at the same church where Maggie had been laid to rest a few months earlier and nearly a year to the day that his book had come out. Although the announcement of his death had not made the newspapers, the church was once again full to overflowing, with farmers, huntsmen, friends and relations, grass verges once more full of cars. The service was taken by the Reverend R D Dew. The service began with Psalm 121 and excerpts from his first poem 'Solitude', read by Malcolm his grandson and another poem of Fred's 'Old Man's Tribute to Lakeland' by Kevin another grandson (Dorothy's sons). There were two more hymns sung, "The day thou gavest, Lord, is ended" and "O Lord my God! When I in awesome wonder." The vicar referred to a passage in St John about knowing your sheep and being able to identify each one individually as Fred could. The Reverend Dew had a copy of Fred's book and mentioned the great love he had for the countryside, animals and people. He finished off his tribute by reading a passage from the introduction and describing Fred as a 'gentleman'.

Words alone cannot describe Fred Nevinson. It has been my privilege to know such a man and one of his belief's I remember so well, being, "If you cannot speak good of a person, don't speak of them at all!"

When we slowly walked out of the church to the graveyard it had been a dull day and it was now after 2 pm. There was a small service again at the grave and a beam of sunlight came down from between the grey clouds. We said our last goodbyes. I looked out across the fields and countryside that Fred so loved and saw a farmer ploughing in a nearby field towards Mealbank. Fred would have enjoyed the view. In a conversation I had previously with his daughter, Dorothy, she said that the one consolation was that, "It was what Dad wanted, to be together again with Mum."

A few weeks later, Clare Hardy, from Tunbridge Wells, in Kent, sent me the latest copy of the Nevinson Newsletter of which she is the editor. This is circulated to Nevinsons all over the world. It contained, on the last page and a half, an article about Fred which had appeared in the Westmorland Gazette in the previous year. Fred would have been so proud of this fitting tribute to him.

I returned to the church and churchyard on Saturday 20 February, 1999, another dull day threatening to rain in the afternoon. I spent some time by the grave, some crocuses and snowdrops had been planted, a gravestone was still awaited and a small light coloured wooden cross marked the grave, bearing the name 'Mr & Mrs Nevinson'. In the last few months we've had a lot of rain and farming is in the doldrums - we have missed him so and that lovely 'twinkle in the eye' with always something nice to say about somebody. Afterwards I went into the church and sat for a while before reading a few of the church magazines. November issue had Fred's poem, 'Ode to an Angel', in it. December had a brief obituary to Fred, with a fuller one in the January 1999 issue. He will never be forgotten, he was the 'Westmorland Shepherd' that so many loved but we have his poems and memories and his family live on. He always said he was never well off financially but was in friends and relations and they are a wealth that cannot be measured.

Today if you visit Skelsmergh Church, a black granite, silver edged headstone marks the last resting place of our beloved 'Westmorland Shepherd', Fred Nevinson and his dear wife, Maggie.

Anne Bonney

Fred being presented with his certificate from Hunter Davis at the 'Book of the Year Awards'
By Courtesy of the Westmorland Gazette